CAMBRIDGE MONOGRAPHS
ON MATHEMATICAL PHYSICS

INTRODUCTION TO SUPERSYMMETRY

To Pauline and Caroline
with love

'Se è giusto che l'immaginazione venga in soccorso alla debolezza della vista, deve essere istantanea e diretta come lo sguardo che l'accende'.*

Italo Calvino, 'Palomar',
Einaudi, Torino, 1983

*'Though it is right for the imagination to come to the rescue of weakness of vision, it must be immediate and direct like the gaze that kindles it.'

INTRODUCTION TO
SUPERSYMMETRY

PETER G.O. FREUND
The University of Chicago, Illinois

CAMBRIDGE UNIVERSITY PRESS
Cambridge
New York Port Chester
Melbourne Sydney

CAMBRIDGE UNIVERSITY PRESS
Cambridge, New York, Melbourne, Madrid, Cape Town, Singapore,
São Paulo, Delhi, Dubai, Tokyo, Mexico City

Cambridge University Press
The Edinburgh Building, Cambridge CB2 8RU, UK

Published in the United States of America by
Cambridge University Press, New York

www.cambridge.org
Information on this title: www.cambridge.org/9780521356756

First published 1986
First paperback edition 1988
Reprinted 1989

A catalogue record for this publication is available from the British Library

Library of Congress Cataloguing in Publication Data

Freund, Peter G.O. (Peter George Oliver), 1936–
Introduction to supersymmetry.
(Cambridge monographs on mathematical physics)
Bibliography
Includes index.
1. Supersymmetry. I. Title. II. Series.
QC174.17.S9F74 1986 530.1′2 85–24308

ISBN 978-0-521-26880-6 Hardback
ISBN 978-0-521-35675-6 Paperback

Contents

Preface

Supersymmetry is one of the boldest, most original and most fruitful ideas to appear in physics in a very long time. In common with nonabelian gauge theories and spontaneous symmetry breaking, it has great depth, and just like those ideas it has travelled quite a way on its own momentum, without carving out for itself a rockbed of supporting experimental evidence. Guided by these analogies, nobody doubts that when discarding some blinding prejudices, or coming by some new data, supersymmetry will come into its own, experimentally as well.

Unlike nonabelian gauge theories and spontaneous symmetry breaking, supersymmetry does not build on well-understood mathematics. Rather, it has created its own, rich, truly new mathematics. Yes, we are faced here with one of those rare instances, when the mathematicians, in all their wisdom, have overlooked a beautiful and most useful structure, and come to appreciate it only at the demand of physicists. We are living in an era in which the contacts between mathematicians and physicists are being vigorously renewed (particularly through work in supersymmetry and gauge theories). This is a good omen, since such contacts have historically always led to great advances both in mathematics and physics.

Supersymmetry is now over a decade old, and excellent monographs and reviews on the subject are available. While teaching a course on supersymmetry at the University of Chicago, it became clear to me that there still was ample space left for a brief introductory text, willing to sacrifice completeness and rigour, in order to achieve a freely flowing exposition of the basic ideas and techniques without all the 'Grassmann clutter'. I have therefore aimed, first of all, at making the mathematics (superspace included) as simple and clear as possible. On the other hand the construction of supersymmetric action principles and the discussion of many of the relevant physical ideas has deliberately been made in low space-dimensionalities where the formulae do not crowd out the concepts. In the process I was forced into some glaring omissions. I said virtually nothing on supersymmetric grand unification, supergraphs, very little on nonrenormalization theorems and on so many more things. As a consolation, the reader will find these topics well presented elsewhere. For what it is worth I included my own assessment of the present status of supersymmetry, of where the field seems to be heading. The table of contents gives an accurate idea of what is – and by implication of what is not – covered in this book.

I have immensely benefited from discussions with Irving Kaplansky, Yoichiro Nambu, Reinhard Oehme, Bruno Zumino, Hermann Nicolai, Tom Curtright, Jeff Rabin, Peter van Nieuwenhuizen, Feza Gürsey, Mike Duff, John Schwarz, Kelly Stelle, Peter West and many others. Phillial Oh and James Wheeler have read the manuscript and made valuable remarks. By thanking my teacher Walter Thirring for his interest in and comments on this book, I can but acquit myself of an infinitesimal fraction of the gratitude I feel towards him.

The writing of this book coincided with a period of deep personal turmoil. The support I received throughout this period from the physics community as a whole, has made me appreciate my good fortune at being one of its members. I was also privileged to enjoy the giving friendship of John and Barbara Ryden, Reinhard and Mafalda Oehme, Elsa Charlston, Stephen Ellis, Subrahmanyan and Lalitha Chandrasekhar, Valentine and Lia Telegdi, Françine Joseph and Joseph Kitagawa. They all command my admiration, respect and gratitude.

Parts of this book were completed during my stays at the Aspen Center for Physics, the Institute for Theoretical Physics at Santa Barbara, and the California Institute of Technology. Their hospitality is gratefully acknowledged, as is the support of the US National Science Foundation.

Lois Cox deserves the ultimate thanks for her technical help, understanding, patience and moral support.

Pasadena, 1985 Peter G. O. Freund

Part I

Supersymmetry: the physical and mathematical foundations

1

From symmetry to supersymmetry

A dynamical system, as studied in physics, is identified by its action integral. In classical physics one extremises this action to find the 'classical paths'. In quantum physics one sums over *all* possible – not only classical – paths, each term weighted by a phase factor $\exp(\mathrm{i}S(\mathrm{path})/\hbar)$, with $S(\mathrm{path})$ the action evaluated at the corresponding path and $\hbar = h/2\pi$ Planck's constant.

In a d-dimensional local field theory the action S is the space–time integral of a local lagrangian density $\mathscr{L}(x)$

$$S = \int \mathrm{d}^d x \mathscr{L}(x)$$

with $\mathscr{L}$ a function of the fields and their first space–time derivatives. Theories of point particles are then also covered, since they can be viewed as field theories in a space–time with one time and zero space dimensions, the coordinates of the point particles playing the role of fields. For concreteness let us consider an example.

The action

$$S = -\int \mathrm{d}^4 x [\partial_\mu \phi^\dagger \partial^\mu \phi + m^2 \phi^\dagger \phi + \lambda(\phi^\dagger \phi)^2] \tag{1.1}$$

specifies a self-interacting complex scalar field ϕ in four-dimensional Minkowski space–time. This action is invariant under the group of Poincaré transformations and under the $U(1)$ group of phase transformation $\phi \rightarrow \mathrm{e}^{\mathrm{i}\alpha}\phi (\alpha = \mathrm{constant})$. According to a theorem of Emmy Noether, to each parameter of the action's symmetry group there corresponds a conserved current (Ramond 1981). From the Poincaré group parameters one gets the energy–momentum tensor and the Lorentz-boost-angular-momentum–density tensor as conserved currents. For the phase transformations, the Noether current is

$$j_\mu = \mathrm{i}\phi^\dagger \overleftrightarrow{\partial}_\mu \phi \equiv \mathrm{i}(\phi^\dagger \partial_\mu - \partial_\mu \phi^\dagger)\phi \tag{1.2}$$

and the conservation law is

$$\partial_\mu j^\mu = 0, \tag{1.3}$$

or in integral form

$$\dot{Q}=0 \quad \text{with} \quad Q=\int_{x^0=\text{const}} d^3x j^0. \tag{1.4}$$

Now in the action (1.1) imagine that ϕ stands not for one complex scalar field but for a column matrix of N such fields

$$\phi=\begin{pmatrix}\phi_1\\ \phi_2\\ \vdots\\ \phi_N\end{pmatrix} \qquad \phi^\dagger=(\phi_1^*,\phi_2^*,\ldots,\phi_N^*).$$

In this case the rephasing group is enlarged to $U(N)$ and there are correspondingly N^2 conserved charges

$$Q_i=\mathrm{i}\int_{x^0=\text{const}} d^3x\phi^\dagger\lambda_i\overleftrightarrow{\partial}_0\phi$$

with λ_i, $i=1,\ldots,N^2$, a basis of hermitean $N\times N$ matrices. So, the more scalar fields in the action, the higher the symmetry. These extra symmetries can be observed both in the free field ($\lambda=0$) and interacting ($\lambda\neq 0$) cases. Were one to achieve the proliferation of fields not by adding ever more scalar fields, but by adding higher spin fields (e.g., vector fields) instead, one could still enlarge the symmetry in the *free* field case. Thus for one free real scalar field ϕ and one free real vector field A_μ both of the *same* mass m, the tensor current

$$j_{\mu\nu}=\phi\overleftrightarrow{\partial}_\mu A_\nu \tag{1.5a}$$

obeys the conservation law

$$\partial^\mu j_{\mu\nu}=\phi\Box A_\nu-\Box\phi A_\nu=(m^2-m^2)\phi A_\nu=0. \tag{1.5b}$$

The tensor $J_{\mu\nu}$ is but one out of an infinity of such conserved currents of the form

$$\phi\overleftrightarrow{\partial}_{\mu_1}\ldots\overleftrightarrow{\partial}_{\mu_n}\phi \quad (n=2,4,6,\ldots), \quad \phi\overleftrightarrow{\partial}_{\mu_1}\ldots\overleftrightarrow{\partial}_{\mu_n}A_\nu \quad (n=1,2,3,\ldots),\ldots \tag{1.6}$$

Just as the charge Q obtained from the vector current was a Lorentz scalar, so the charge $Q_\nu=\int_{x^0=\text{const}} j_{0\nu}d^3x$ obtained from $j_{\mu\nu}$ is a Lorentz vector, as is $\int_{x^0=\text{const}}\phi\overleftrightarrow{\partial}_0\overleftrightarrow{\partial}_\nu\phi d^3x$, whereas the (integral) charges obtained from the remaining currents (1.6) have ever higher tensorial rank. By analogy with the vector current above, one may ask whether these extra conservation laws can also be extended to the interacting case.

In dimension $d > 2$ the answer to this question (Coleman & Mandula 1967, Witten 1981) turns out to be an emphatic no! Specifically, in dimension $d > 2$, consider a relativistic quantum theory with a discrete spectrum of massive (but no massless) one-particle states, and with some nonzero scattering amplitudes. In any such theory, as we shall argue below, the only conserved tensorial charges that are not Lorentz scalars, are the energy–momentum vector P_μ and the angular-momentum–Lorentz-boost tensor $M_{\mu\nu}$ which span the Poincaré algebra $\mathscr{P}$. All other conserved charges in such a theory must be Lorentz scalars. As a rule they span a compact 'internal symmetry' Lie algebra (which thus commutes with $\mathscr{P}$). The total symmetry algebra is thus *always* of the form

$$\mathscr{P} + \mathscr{g}.$$

In the totally massless case, $\mathscr{P}$ can be extended to the conformal algebra $\mathscr{K}$ (see Mack & Salam 1969). To see how this result comes about, consider a forbidden tensorial charge, say a conserved second rank (hermitean) tensor charge $Q_{\alpha\beta}$ which for simplicity we shall assume traceless $Q^\alpha_\alpha = 0$. Assume a scalar particle of mass m, carrying the charge $Q_{\alpha\beta}$, appears in the theory. Let $|p>$ be a corresponding one-particle state, $p^2 = -m^2$. Then the expectation value $\langle p|Q_{\alpha\beta}|p\rangle$ is of the form ($\eta_{\alpha\beta} = \mathrm{diag}(-1, +1, +1, \ldots, +1)$ is the d-dimensional Minkowski metric tensor)

$$\langle p|Q_{\alpha\beta}|p\rangle = \left(p_\alpha p_\beta - \frac{1}{d}\eta_{\alpha\beta}p^2\right)C,$$

where $C \neq 0$ is a real number. Consider now the process whereby two such incoming particles of momenta p_1, p_2 scatter and then go out with final momenta p'_1 and p'_2. The conservation law of $Q_{\alpha\beta}$ applied between asymptotic incoming and outgoing states requires

$$C\left[p_{1\alpha}p_{1\beta} + p_{2\alpha}p_{2\beta} + \frac{1}{d}\eta_{\alpha\beta}(m^2 + m^2)\right]$$
$$= C\left[p'_{1\alpha}p'_{1\beta} + p'_{2\alpha}p'_{2\beta} + \frac{1}{d}\eta_{\alpha\beta}(m^2 + m^2)\right],$$

where α, $\beta = 1, \ldots, d$. If $C \neq 0$, these equations imply that the scattering must proceed either in the forward or backward direction; (e.g., the $\alpha = 0$, $\beta = 0$ equation requires the sums of the squares of the particle energies to be the same in the initial and final states) whereas in all other directions there can be no scattering. This conflicts with the known analyticity properties of scattering amplitudes (in dimension $d > 2$!) and thus excludes

$C \neq 0$, so that no interacting scalar particle can carry this charge. Similar arguments when the nonvanishing amplitudes involve other than identical scalar particles can then be used to rigorously prove this 'no-go theorem'. In two dimensions this type of argument does not apply, since forward and backward scattering is all one can have there. Indeed, nontrivial completely integrable quantum systems in two dimensions are known, in which infinite towers of charges of ever higher tensorial rank are all conserved (see for example, Berg, Karowski & Thun, 1976).

These arguments establish the impossibility of nontrivial symmetries that connect particles of different spins, if *all* these particles have integer spin, or if *all* have half-odd-integer spin. Were the symmetry to connect particles of integer spin with particles of half-odd-integer spin, then some of the charges would turn out to be spinorial and the 'no-go theorem' would not apply. The change is quite drastic. Indeed, tensorial charges are space integrals of *tensorial* local fields (see for example, equation (1.4) for the charge Q). Spinorial charges should then also be given as space integrals of spinorial local fields. But the famous spin–statistics connection of quantum field theory (Streater & Wightman, 1964) instructs us that at space-like distances, local spinor fields *cannot* commute, rather they must anticommute. This leads us to expect that the commutator of two spinorial charges will *not* be determined, so that these charges along with the tensorial charges cannot span an ordinary Lie algebra. For two spinorial charges the spin–statistics connection dictates that the anticommutator be determined, and thus one is led to invent a new type of Lie-like algebra in which the bracket [,] is not always an antisymmetric operation (like a commutator) but 'occasionally' a symmetric operation (like an anti-commutator). Such algebras are called *Lie superalgebras* and will be described and classified in the next chapter.

2

Lie superalgebras

An ordinary Lie algebra $\mathscr{g}$ over the field of complex numbers $\mathbb{C}$ is specified by the following three axioms (Humphreys 1972):

(i) $\mathscr{g}$ is a vector space over $\mathbb{C}$.

(ii) $\mathscr{g}$ is endowed with a binary operation, the bracket [,], which is bilinear and anticommutative.

In detail, to any two elements A and B of $\mathscr{g}$ is attached the bracket $[A,B]$ such that

$$[aA + a'A', B] = a[A,B] + a'[A',B] \quad \text{for } a \text{ and } a' \in \mathbb{C}$$

and

$$[A,B] = -[B,A].$$

(iii) The bracket operation obeys the Jacobi identity

$$[A,[B,C]] + [B,[C,A]] + [C,[A,B]] = 0.$$

To get to a Lie superalgebra[†] we have to deal with a vector space which has two types of elements: Bose and Fermi. Thus we use a *graded* vector space such that each of its vectors has a grade defined mod 2. We call Bose the even, or grade 0 elements, and Fermi the odd, or grade 1 elements of this graded vector space. The set ${}^0V({}^1V)$ of all Bose (Fermi) vectors is a vector space itself. ${}^0V$ and ${}^1V$ have only one element in common: the null element. Now the problem arises what grade is one to give to a linear combination of a Bose with a Fermi element. In physics it is known that superselection rules (Wick, Wightman & Wigner 1952) forbid the consideration of such combinations. Mathematically we express this by requiring the total vector space V to be the set-theoretic union of ${}^0V$ and ${}^1V$: $V = {}^0V \cup {}^1V$, rather than their direct sum ${}^0V + {}^1V$. The axioms of a Lie superalgebra $\mathscr{s}$ are then

(i) *$\mathscr{s}$ is a mod 2 graded vector space over* $\mathbb{C}$.

[†] We refrain from the use of the term graded Lie algebra (GLA) encountered in the early literature, since ordinary Lie algebras can also be graded and confusion can arise.

(ii) $\mathscr{s}$ *is endowed with a binary operation, the bracket, which is bilinear, superanticommutative and mod 2 grade additive.*

In detail this means that to any pair A, B of elements of $\mathscr{s}$ we associate the bracket $[A, B]$ which while still bilinear as in the ordinary case, is no longer anticommutative. Rather $[A, B] = -[B, A]$ in all cases but one, namely when both A and B are Fermi in which case $[A, B] = +[B, A]$. Let us denote by a, b, c,... the grades, valued in the set $(0, 1)$, of the elements A, B, C,... of $\mathscr{s}$. Mod 2 grade additivity means $[A, B] = C \rightarrow a + b = c \pmod 2$. Thus the bracket of two Bose or of two Fermi elements is Bose, whereas the bracket of a Bose and a Fermi element is Fermi, as expected.

(iii) *The bracket operation obeys the super-Jacobi identity*

$$(-1)^{ac}[A,[B,C]] + (-1)^{ba}[B,[C,A]] + (-1)^{cb}[C,[A,B]] = 0$$

This reduces to the ordinary Jacobi identity in all cases but one: when any two of the elements A, B, C are Fermi and the third one is Bose in which case one of the three usual Jacobi terms has its sign flipped.

Important notation: On account of its superanticommutativity, the bracket of a Lie superalgebra if realized as a commutator when one or both bracketed elements are Bose, becomes an anticommutator when both bracketed elements are Fermi. The standard notation for these operations is $[,]$ and $\{,\}$. We shall nevertheless use a common notation $[,]$ for all these cases, since given the elements to be bracketed, their grades unambiguously specify the nature of the bracket (commutative, or anticommutative). On rare occasions, to emphasize the commutativity properties of a particular bracket under consideration, we will, redundantly, use the notations $[,]_+$, $[,]_-$ for anticommutators and commutators respectively. Many other notational systems are to be found in the literature, all the way from spelling things out case by case and using ordinary $[,]$, $\{,\}$ notation, to using the – in my opinion, weird – notation $[,\}$ for a superanticommutative bracket.

For ordinary Lie algebras the 'building blocks' are the simple ones fully classified by Cartan and Killing (Humphreys 1972). For Lie superalgebras, although they had occurred in various mathematical contexts (Fröhlicher & Nijenhuis 1956, Gerstenhaber 1963, 1964, Milnor & Moore 1965), the classification problem (of the simple ones) was not addressed until after their appearance in physics. A Lie superalgebra $\mathscr{s}$ is *simple* if it has no nontrivial ideals (in other words, any subsuperalgebra ℓ of $\mathscr{s}$, such that $L \in \ell$ and $A \in \mathscr{s}$ always implies $[L, A] \in \ell$, is trivial, i.e., either $\ell = 0$ or $\ell = s$).

The simple finite-dimensional Lie superalgebras over $\mathbb{C}$ are now fully classified (Kac 1975, 1977; Freund & Kaplansky 1976; Nahm & Scheunert 1976; Scheunert 1979; Kaplansky 1980). There are eight infinite families $s\ell(m|n)$, $osp(m|n)$, $P(n)$, $Q(n)$, $W(n)$, $S(n+2)$, $\tilde{S}(n+2)$, $H(n+3)$, a continuum $D(2|1;\alpha)$ of 17-dimensional exceptional superalgebras, and one exceptional superalgebra each in dimensions 31 and 40 ($G(3)$ and $F(4)$ respectively). We shall now describe these superalgebras although for physics the relevant ones are, above all, the special linear and the orthosymplectic ones.

The special linear superalgebras $s\ell(m|n)$

Consider a mod 2 graded vector space $V(m|n)$ over $\mathbb{C}$ with m Bose (or even) and n Fermi (or odd) dimensions. Represent a vector in $V(m|n)$ as a column matrix with $m+n$ rows. A Bose (Fermi) element will thus have nonzero entries in the top m (bottom n) entries of this column matrix. Now consider the complex linear transformations on $V(m|n)$. The grading of $V(m|n)$ induces an obvious grading of these linear transformations. In the matrix representation a Bose linear transformation (that carries Bose vectors into Bose vectors and Fermi vectors into Fermi vectors) is block diagonal

$$\begin{array}{c} \\ m \\ n \end{array}\begin{array}{c} \begin{array}{cc} m & n \end{array} \\ \left(\begin{array}{c|c} \square & 0 \\ \hline 0 & \square \end{array}\right) \end{array}$$

whereas a Fermi transformation is block off-diagonal

$$\begin{array}{c} \\ m \\ n \end{array}\begin{array}{c} \begin{array}{cc} m & n \end{array} \\ \left(\begin{array}{c|c} 0 & \square \\ \hline \square & 0 \end{array}\right) \end{array}$$

In the matrix representation the bracket is the usual commutator in all cases but one, namely when both bracketed transformations are Fermi in which case it is an anticommutator. It is obvious that one obtains a Lie superalgebra in this way, but, like in the ordinary case, this general linear algebra $g\ell(m|n)$ is *not* simple. In the ordinary case one achieves simplicity by imposing the tracelessness condition, the ordinary bracket (commutator) of any (finite-dimensional) matrices being always traceless. For two Fermi transformations in $\mathscr{gl}(m|n)$ the bracket is now an anticommutator, and even though the individual Fermi transformations are traceless (in the ordinary sense) their anticommutator, in general, is *not*. We therefore have to invent a new concept of trace, call it *supertrace*, such, that it identically vanishes for the superalgebra bracketing of two matrices. It is readily checked that for a matrix

$$M = \begin{matrix} & m & n \\ m & A & B \\ n & C & D \end{matrix}$$

The supertrace is given by

$$\operatorname{str} M = \operatorname{tr} A - \operatorname{tr} D \tag{2.1}$$

which differs from the ordinary trace by the relative sign of the two terms.

To achieve simplicity we restrict ourselves to the supertraceless elements of $\mathscr{gl}(m|n)$. For $m \neq n$ the supertraceless $(m+n) \times (m+n)$ matrices form a *simple* $((m+n)^2 - 1)$-dimensional superalgebra $\mathscr{sl}(m|n)$ under the bracketing described above.

In the '*balanced*' case when $m = n$, the unit matrix is supertraceless, so that even after the imposition of supertracelessness the superalgebra $\mathscr{sl}(m|m)$ still has a one-dimensional center (the unit matrix commutes with all elements of $\mathscr{sl}(m|m)$) which has to be divided out. In this case the simple superalgebra $\mathscr{Psl}(m|m)$ with $m \geqslant 2$ has dimension $4m^2 - 2$ ($\mathscr{Psl}(1|1)$ is nilpotent).

The Bose (even) sectors ${}^0\mathscr{s}$ of a Lie Superalgebra $\mathscr{s}$ is an ordinary Lie algebra. In general, this ordinary Lie algebra will not be simple but will contain a piece that shuffles only the bosons, a piece that shuffles only the fermions, and a piece that shuffles the bosons among themselves and the fermions among themselves, but in a correlated manner. Thus

$${}^0\mathscr{gl}(m|n) = \mathscr{gl}(m) + \mathscr{gl}(n)$$
$${}^0\mathscr{sl}(m|n)|_{m\neq n} = \mathscr{sl}(m) + \mathscr{sl}(n) + \text{one-dimensional abelian piece}$$
$${}^0\mathscr{Psl}(m|m) = \mathscr{sl}(m) + \mathscr{sl}(m)$$

We thus recognize the Bose parts of $\mathscr{gl}(m|n)$, $\mathscr{sl}(m|n)|_{m\neq n}$, $\mathscr{Psl}(m|m)$ to have dimensions $m^2 + n^2$, $m^2 + n^2 - 1$, $2m^2 - 2$, respectively. The dimension of the corresponding Fermi parts are then $2mn$, $2mn$, $2m^2$.

The orthosymplectic superalgebras $osp(m|n)$

Choose n even and endow the graded vector space $V(m|n)$ with a bilinear form

$$(x, y) \equiv x^{\mathrm{T}} G y$$

which is symmetric (antisymmetric) on the Bose (Fermi) sector of $V(m|n)$

$$G = \left(\begin{array}{ccc|ccccc} 1 & & & & & & & \\ & 1 & \vdots & & & & & \\ & & 1 & & & & & \\ \hline & & & & 1 & & & \\ & & & -1 & & 1 & & \\ & & & & -1 & & \ddots & \\ & & & & & & & 1 \\ & & & & & & -1 & \end{array}\right) \begin{array}{l} \\ m \\ \\ \\ \\ n \\ \\ \\ \end{array}$$

(column blocks: m, n)

Now consider those complex linear transformations U on $V(m|n)$ on which we impose a 'superantisymmetry' condition. Naively one could choose this condition as

$$(x, Uy) + (Ux, y) = 0,$$

but, on account of the different ordering of U and x in the two terms, one has to correct with a minus sign in the Fermi–Fermi case. To define $osp(m|n)$ we thus impose (u, x in the exponent are the grades of U, x, as defined above)

$$(x, Uy) + (-1)^{ux}(Ux, y) = 0$$

The Bose sector of $osp(m|n)$ is then obviously $o(m) + sp(n)$ ($n = $ even!), and as such has dimension $\frac{1}{2}m(m-1) + \frac{1}{2}n(n+1)$. To determine the Fermi sector we choose a Bose vector $b \in V(m|n)$, a Fermi vector $f \in V(m|n)$ and a Fermi $U \in osp(m|n)$. For $x = b$, $y = f$, and such a Fermi U, the superantisymmetry condition reduces to

$$b^{\mathrm{T}} G U f + b^{\mathrm{T}} U^{\mathrm{T}} G f = 0$$

which determines

$$U = \begin{pmatrix} 0 & V \\ W & 0 \end{pmatrix} \begin{array}{l} m \\ n \end{array} \quad V = -W^T C \quad C = \begin{pmatrix} & 1 & & & & \\ -1 & & 1 & & & \\ & -1 & & 1 & & \\ & & -1 & & \ddots & \\ & & & \ddots & & 1 \\ & & & & -1 & \end{pmatrix}$$

(column blocks of U: m, n)

For $x=f$ and $y=b$ one obtains the transpose of the same condition. Thus the lower half of the Fermi matrix U determines its upper half (or vice versa). There are thus only mn fermionic generators to $\mathscr{osp}(m|n)$ (not $2mn$ as for $\mathscr{sl}(m|n)$). The total dimension of $\mathscr{osp}(m|n)$ is $\frac{1}{2}[(n+m)^2+n-m]$. Had we chosen G symmetric in both the Fermi and Bose sectors or antisymmetric in both, the Fermi sectors of the resulting algebras would have been empty and we would have 'rediscovered' the *ordinary* Lie algebras $\mathscr{o}(m+n)$, $\mathscr{sp}(m+n)$. If for m even we had chosen G anti-symmetric over the Bose sector and symmetric over the Fermi sector we would have obtained $\mathscr{osp}(n|m)$. So, we only find something new for a G that has both a symmetric and an antisymmetric part, though it is mathematically irrelevant over which of the two sectors of $V(m|n)$ we impose which symmetry of the metric.

The superalgebras $P(n)$

$P(n)$ is the subsuperalgebra of $\mathscr{sl}(n+1|n+1)$ defined by the matrices of the form

$$\begin{pmatrix} a & b \\ c & -a^{\mathrm{T}} \end{pmatrix} \quad \text{with } \operatorname{tr} a = 0, b = b^{\mathrm{T}}, c = -c^{\mathrm{T}}$$

The dimension of $P(n)$ is $2(n+1)^2-1$. These superalgebras have been first considered by Gell-Mann (see Gell-Mann & Ne'eman, 1964) and by Michel and Radicati (see Michel, 1969 and also Corwin, Ne'eman & Sternberg, 1975).

The superalgebras $Q(n)$

$Q(n)$ is the $[2(n+1)^2-2]$-dimensional subsuperalgebra of $\mathscr{sl}(n+1|n+1)$ defined by the matrices of the form $\begin{pmatrix} a & b \\ b & a \end{pmatrix}$ after one divides out the center corresponding to the unit matrix and imposes $\operatorname{tr} b = 0$.

Exceptional Lie superalgebras

Just as in the ordinary case, there are some superalgebras which only exist for a particular dimensionality. They are called *exceptional*. The lowest exceptional superalgebra is a continuously infinite set of 17-dimensional algebras $D(2|1;\alpha)$ labeled by $\alpha \in \mathbb{R}$. They all have Bose sector $\mathscr{o}(4)+\mathscr{sp}(2)$, and an eight-dimensional Fermi sector which transforms like $(\mathbf{4},\mathbf{2})$ under $\mathscr{o}(4)+\mathscr{sp}(2)$. Next the 31-dimensional superalgebra $G(3)$ has $\mathscr{g}_2+\mathscr{su}(2)$

in its Bose sector. Its 14-dimensional Fermi sector transforms like the $(\mathbf{7}, \mathbf{2})$ representation of $g_2 + su(2)$. This is the only time any of the exceptional ordinary Lie algebras makes an appearance in superalgebra theory.

Finally, $F(4)$ is 40-dimensional. Its Bose sector is $spin(7) + su(2)$. Its Fermi sector transforms like the $(\mathbf{8}, \mathbf{2})$ representation of $spin(7) + su(2)$. Here $\mathbf{8}$ is the eight-dimensional spinorial representation of $spin(7)$. This superalgebra plays the role of an extended de-Sitter algebra in a six-dimensional space–time (Nahm 1978).

Superalgebras of Cartan type

Consider a n-dimensional space with local coordinates x^μ $\mu = 1, 2, \ldots, n$. One of the ordinary infinite simple Lie algebras of Cartan type is that of analytic coordinate transformations in such a space. It is (locally) generated by the infinite set of generators.

$$(x^1)^{m_1} \ldots (x^n)^{m_n} \partial_\mu \quad m_i = 0, 1, \ldots . \tag{2.2}$$

Now assume x^μ not to be ordinary real numbers but generators of a Grassmann algebra

$$x^\mu x^\nu + x^\nu x^\mu = 0$$
$$\partial_\mu x^\nu = \delta^\nu_\mu .$$

Then the set of generators of type (2.2) truncates since each m_i can only take two values, 0 and 1. One thus obtains a Lie superalgebra $W(n)$ of *finite* dimension $n2^n$: $W(n)$ is then the algebra of analytic coordinate transformations on a n-dimensional 'fermionic space', or more mathematically, the algebra of derivations of the Grassman algebra with n generators. $W(n)$ is simple for $n \geqslant 2$. There are three more families of finite-dimensional Lie superalgebras of Cartan type $S(n+1)$, $\tilde{S}(n+2)$ and $H(n+2)$, which all are simple for $n \geqslant 2$, but which we shall not describe here in detail (see Kac 1977).

No matter what the space–time dimension, the ordinary Poincaré algebra is not even semisimple, let alone simple. So one may question the wisdom of paying this much attention to *simple* Lie superalgebras. Yet on the one hand, the Poincaré algebra can be obtained by Wigner–Inönü contraction (to be described in chapter 4) from the simple de-Sitter algebra and, on the other hand, it can be embedded in the again simple conformal algebra. We shall see that similar things happen for the 'Poincaré superalgebras' as well, and thereby simple Lie superalgebras will come to play a central role. The selection of the physically relevant simple superalgebras is achieved using two principles: compatibility with the spin–statistics connection and circumvention of the no-go theorem of chapter 1 (Haag, Łopuszanski & Sohnius 1975). We shall come to identify the physically relevant algebras in chapter 4. For the time being we still lack a technical ingredient: the theory of spinors in d-dimensional (euclidean or minkowskian) space. We therefore devote the next chapter to this topic.

span the ordinary Lie algebra *spin*$(p, q; \mathbb{R})$ if one identifies Lie bracketing with commutation. A representation of the Clifford algebra thus yields a representation of *spin*$(p, q; \mathbb{R})$. The elements of the corresponding representation space are called *spinors*.

It is a classical result (Chevalley 1954) that a Clifford algebra admits an irreducible matrix representation, unique up to equivalence (for Dirac's Clifford algebra the proof of this statement is given in many physics textbooks). It is therefore possible to uniquely characterize all Clifford algebras as matrix algebras, as we shall now do (Atiyah, Bott, & Shapiro 1964, Coquereaux 1982).

We first study the Clifford algebras for $\mathbb{F} = \mathbb{R}$. Call $C(p, q)$ the Clifford algebra corresponding to the quadratic form (3.3). We shall show that knowledge of, say, $C(1,0), C(2,0), \ldots, C(8,0)$ determines every other $C(p, q)$, and that the interesting properties of $C(p, q)$ depend only on the signature $(p-q)$ mod 8. Let us start by considering the first few $C(p, q)$s.

C(1,0): e_1 is the one-dimensional unit matrix $\mathbf{1}_1$ and hence from (3.2) the general element of $C(1,0)$ is $a_{(0)} + a_{(1)} \cdot \mathbf{1}_1$ with $a_{(0)}, a_{(1)} \in \mathbb{R}$ so that $C(1,0)$ is $\mathbb{R} + \mathbb{R}$.

C(0,1): $e_1 = i\mathbf{1}_1$ so that from (3.2) the general element of $C(0,1)$ is of the form $a_{(0)} + a_{(1)} i\mathbf{1}_1$, i.e., $C(0,1) = \mathbb{C}$.

C(2,0): $e_1 = \sigma_1, e_2 = \sigma_3, e_1^2 = e_2^2 = \mathbf{1}_2, e_1 e_2 = -i\sigma_2$ are a basis of the algebra $\mathbb{R}(2)$ of real 2×2 matrices $C(2,0) = \mathbb{R}(2)$

C(1,1): $e_1 = \sigma_1$, $e_2 = i\sigma_2$, $e_1^2 = -e_2^2 = \mathbf{1}_2$, $e_1 e_2 = -\sigma_3$ again imply $C(1,1) = \mathbb{R}(2)$.

C(0,2): $e_1 = i\sigma_1$, $e_2 = i\sigma_2$, $e_1 e_2 = -i\sigma_3$, $e_1^2 = e_2^2 = -\mathbf{1}_2$ is a basis of quaternion algebra $\mathbb{H}$, so that $C(0,2) = \mathbb{H}$.

We may add that $C(0,0)$ is but the algebra of the reals $C(0,0) = \mathbb{R}$ since there are no e_is. We thus have

$$\left.\begin{array}{lll} C(0,0) = \mathbb{R} & C(1,0) = \mathbb{R} + \mathbb{R} & C(0,1) = \mathbb{C} \\ C(2,0) = \mathbb{R}(2) & C(1,1) = \mathbb{R}(2) & C(0,2) = \mathbb{H} \end{array}\right\} \tag{3.4}$$

From these simple building blocks we can construct all the other $C(p, q)$s. A straightforward generalization of the familiar presentation of the algebra of four-dimensional Dirac matrices as a direct product of two copies of the Pauli algebra yields the isomorphisms:

$$\left.\begin{array}{l} C(p,q) \otimes C(2,0) \sim C(q+2, p) \\ C(p,q) \otimes C(1,1) \sim C((p+1), q+1) \\ C(p,q) \otimes C(0,2) \sim C(q, p+2) \end{array}\right\} \tag{3.5}$$

Finally, if we call $\mathbb{K}(n)$ the $n \times n$ matrix algebra over the field $\mathbb{K} = \mathbb{R}, \mathbb{C}, \mathbb{H}$,

3
Spinors in d-dimensions

Consider a d-dimensional vector space V over the field $\mathbb{F}$, for which we shall choose the two alternatives $\mathbb{F}=\mathbb{R}$ and $\mathbb{F}=\mathbb{C}$ ($\mathbb{R}$ = field of real numbers, $\mathbb{C}$ = field of complex numbers). Let Q be a quadratic form on V.

$$Q: x \in V \rightarrow Q(x) \in \mathbb{F}.$$

This defines a symmetric scalar product on V: to any pair $x \in V$, $y \in V$ we associate the scalar product

$$\phi(x, y) \equiv xy + yx = Q(x+y) - Q(x) - Q(y)$$

In particular, for e_μ, $\mu = 1, \ldots, d$, a basis of V, orthogonal with respect to Q we then have

$$e_\mu e_\nu + e_\nu e_\mu = 2\delta_{\mu\nu} Q(e_\mu) \cdot \mathbf{1} \tag{3.1}$$

The associative algebra with unit element generated by the e_μ with the defining relations (3.1) is called the *Clifford algebra* $C(Q)$ of the quadratic form Q. In particular, the Clifford algebra for the identically null quadratic form $Q(x) \equiv 0$, is the Grassman algebra encountered in chapter 2. Here, however, we shall pursue nondegenerate quadratic forms. The dimension of $C(Q)$ is 2^d. A convenient basis for $C(Q)$ is

$$1, e_\mu, e_{\mu_1}e_{\mu_2}, e_{\mu_1}e_{\mu_2}e_{\mu_3}, \ldots, e_{\mu_1}e_{\mu_2}\ldots e_{\mu_d} \quad \text{with } \mu_1 < \mu_2 < \mu_3 < \cdots < \mu_d.$$

The general element of $C(Q)$ can be written in this basis as

$$a_{(0)} + a^{\mu}_{(1)} e_\mu + a^{\mu\nu}_{(2)} e_\mu e_\nu + \cdots + a^{1\cdots\mu_d}_{(d)} e_{\mu_1} \cdots e_{\mu_d} \tag{3.2}$$

with the $\mathbb{F}$-valued coefficients $a^{\mu_1\cdots\mu_i}_{(i)}$ totally antisymmetric in the Greek indices. Now choose $\mathbb{F}=\mathbb{R}$ and specify the quadratic form by

$$Q(e_\mu) = \begin{cases} +1 & \mu = 1, 2, \ldots, p \\ -1 & \mu = p+1, \ldots, p+q = d. \end{cases} \tag{3.3}$$

Then the elements (3.2) of the form

$$\tfrac{1}{2} a^{\mu\nu}(e_\mu e_\nu - e_\nu e_\mu)$$

we note the isomorphisms:

$$\left.\begin{array}{l}\mathbb{R}(m)\otimes\mathbb{R}(n)\sim\mathbb{R}(mn)\\ \mathbb{R}(1)\sim\mathbb{R}\\ \mathbb{R}(m)\otimes\mathbb{C}\sim\mathbb{C}(m)\\ \mathbb{R}(m)\otimes\mathbb{H}\sim\mathbb{H}(m)\\ \mathbb{C}\otimes\mathbb{C}\sim\mathbb{C}\oplus\mathbb{C}\\ \mathbb{H}\otimes\mathbb{C}\sim\mathbb{C}(2)\\ \mathbb{H}\otimes\mathbb{H}\sim\mathbb{R}(4)\end{array}\right\} \tag{3.6}$$

Here $\mathbb{C}$ ($\mathbb{H}$) is again viewed as a 2- (4-) dimensional algebra over the reals. Using the isomorphisms (3.4)–(3.6) one readily constructs $C(n,0)$ and $C(0,n)$ for $0 \leqslant n \leqslant 8$ with the results shown in Table 3.1. Repeated use of the isomorphisms (3.5) along with the isomorphism

$$C(0,2)\times C(0,2)\times C(2,0)\times C(2,0)=\mathbb{R}(16)=C(0,8)=C(8,0)$$

yields

$$C(0,n+8)\sim C(0,n)\otimes C(0,8)$$
$$C(n+8,0)\sim C(n,0)\otimes C(8,0).$$

Both $C(0,8)$ and $C(8,0)$ being $\mathbb{R}(16)$, this means by (3.6) that

$$C(0,n)=\mathbb{K}(l)\rightarrow C(0,n+8)=\mathbb{K}(16l) \tag{3.7}$$

and a similar formula for $C(n+8,0)$. Thus from the table 3.1 we can calculate $C(n,0)$ and $C(0,n)$ for all n. For instance

$$C(0,137)=C(0,1+(17\times 8))=\mathbb{C}(16^{17})$$

A less facetious example is come by, comparing the $n=0$ and $n=8$ entries in Table 3.1. So far we have only determined $C(n,0)$ and $C(0,n)$ for all n. But any $C(p,q)$ can be obtained according to (3.5) either from

Table 3.1

n	$C(n,0)$	$C(0,n)$
0	$\mathbb{R}$	$\mathbb{R}$
1	$\mathbb{R}+\mathbb{R}$	$\mathbb{C}$
2	$\mathbb{R}(2)$	$\mathbb{H}$
3	$\mathbb{C}$	$\mathbb{H}+\mathbb{H}$
4	$\mathbb{H}(2)$	$\mathbb{H}(2)$
5	$\mathbb{H}(2)\oplus\mathbb{H}(2)$	$\mathbb{C}(4)$
6	$\mathbb{H}(4)$	$\mathbb{R}(8)$
7	$\mathbb{C}(8)$	$\mathbb{R}(8)\oplus\mathbb{R}(8)$
8	$\mathbb{R}(16)$	$\mathbb{R}(16)$

$C(p-q,0)$ or $C(0,q-p)$ by repeated multiplication with $C(1,1)$. The equation $C(1,1)=\mathbb{R}(2)$ and (3.6) then allow the determination of all $C(p,q)$ from the already known prototype $C(p-q,0)$ or $C(0,q-p)$. The results are presented in Table 3.2. Finally, we are interested in the complex Clifford algebras. These are trivially obtained by complexifying the real $C(p,q)$. Obviously the complexification of all $C(p,q)$ with the same $p+q$ will be the same. Let $\bar{C}(d)=C(p,d-p)\times\mathbb{C}$ be the complexification of $C(p,d-p)$; then we have table 3.3. This is even simpler than table 3.2 for the real Clifford algebras. With the classification of Clifford algebras complete, we are now in measure to discuss the various types of spinors: Dirac, Majorana, Weyl, Majorana–Weyl.

The elements of the representation space of $\bar{C}(d)$ are called (complex) *Dirac spinors.* From table 3.3 it is clear that Dirac spinors exist in any dimension and that a Dirac spinor has $2^{[\frac{1}{2}d]}$ complex components among which one can impose, for instance, the Dirac equation.

If there exists a real representation of the Clifford algebra corresponding

Table 3.2

$p-q$ (mod 8)	$C(p,q)$
0	$\mathbb{R}(2^l)$
1	$\mathbb{R}(2^l)\oplus\mathbb{R}(2^l)$
2	$\mathbb{R}(2^l)$
3	$\mathbb{C}(2^l)$
4	$\mathbb{H}(2^{l-1})$
5	$\mathbb{H}(2^{l-1})\oplus\mathbb{H}(2^{l-1})$
6	$\mathbb{H}(2^{l-1})$
7	$\mathbb{C}(2^l)$

In this table $l=[d/2]$ is the integer part of $d/2$ and $d=p+q$.

Table 3.3

d(mod 2)	$\bar{C}(d)$
0	$\mathbb{C}(2^{[\frac{1}{2}d]})$
1	$\mathbb{C}(2^{[\frac{1}{2}d]})+\mathbb{C}(2^{[\frac{1}{2}d]})$

to a given space–time dimension and metric signature, then the (massive or massless) Dirac equation will shuffle the real parts of any Dirac spinor among themselves and the imaginary parts among themselves. It is then possible to impose a reality condition on spinors, say that the spinor be real, without creating a contradiction with Dirac's equation. In those dimensions and metric signatures where such a real representation of the Clifford algebra exists we can then have real or *Majorana spinors.* From table 3.2 this is possible in any dimension provided only

- M: the metric signature $p-q=0$, 1, or 2 (mod 8).

In the case of the massless Dirac equation a reality condition on spinors is possible even if the Clifford algebra is not real, provided all its generators (γ-matrices) are pure imaginary (the overall i then factors out and as far as the Dirac equation is concerned, everything proceeds as if there existed a real representation). A pure imaginary (or pseudo-Majorana) representation of the Clifford generators of $C(p,q)$ will always exist if $C(q,p)$ has a pure real representation which according to condition M above means $q-p=0$, 1, 2 (mod 8) or equivalently $p-q=0$, 6, 7 (mod 8). So the condition for the existence of a pseudo-Majorana representation is

- M′: the metric signature $p-q=0$, 6, 7 (mod 8).

For any Clifford algebra $C(p-q)$ (or $\bar{C}(d)$) the elements of form (3.2) with only even terms, i.e., with $a^{\mu}_{(1)} = a^{\mu\nu\rho}_{(3)} = \cdots = 0$, form a (2^{d-1})-dimensional subalgebra ${}^{0}C(p,q)$ (or ${}^{0}\bar{C}(d)$). This algebra is not simple, in general. This is seen by considering the Clifford algebra element

$$\varepsilon = e_1 e_2 e_3 \ldots e_d,$$

the generalization to d-dimensions of the familiar γ_5 of Dirac. It is readily checked that

$$\varepsilon^2 = \begin{cases} +1 & \text{for } p-q=0 \text{ or } 1 \pmod 4) \\ -1 & \text{for } p-q \neq 0 \text{ or } 1 \pmod 4). \end{cases}$$

When $\varepsilon^2 = +1$ then, the analogues of $\frac{1}{2}(1 \pm \gamma_5)$,

$$P_{\pm} = \tfrac{1}{2}(1 \pm \varepsilon)$$

are projection operators.

For complex Clifford algebras $\bar{C}(d)$ the condition $\varepsilon^2 = +1$ is not necessary, since for $\varepsilon^2 = -1$, the now legal iε, whose square is $+1$, is available for the construction of $P_{\pm}$. Thus for even d (so that $\varepsilon \in {}^{0}\bar{C}(d)$) the complex Clifford algebra ${}^{0}\bar{C}(d)$ is not simple, it falls apart into two simple ideals.

$$^0\bar{C}(d) = {}^0\bar{C}(d)P_+ + {}^0\bar{C}(d)P_-$$

with $P_\pm$ defined as $\frac{1}{2}(1 \pm \varepsilon)$ or $\frac{1}{2}(1 \pm i\varepsilon)$ depending on whether ε or $i\varepsilon$ squares to one. The full Lie algebra $\mathit{spin}(p, q; R)$ is contained in $^0\bar{C}(d)$ $(d = p + q)$, so that for all even d the complex Dirac spinor representation is reducible into two complex *Weyl spinors*. We thus have complex Weyl spinors whenever

■ W: d is even

Retreating now to real Clifford algebras, $^0C(p, q)$ with even $d = p + q$ will split in a similar pattern

$$^0C(p, q) = {}^0C(p, q)P_+ + {}^0C(p, q)P_-$$

whenever $p - q = 0 \pmod 4$ (so that $\varepsilon^2 = +1$, $i\varepsilon$ is now unavailable). In particular, this implies that Majorana spinors can be further reduced into Majorana–Weyl spinors whenever simultaneously $p - q = 0, 2 \pmod 8$ *and* $(p - q) = 0 \pmod 4$. In other words, we have Majorana–Weyl spinors whenever

■ $\bowtie$: $p - q = 0 \pmod 8$.

Note that in these $\bowtie$ dimensions one has both Majorana and pseudo-Majorana spinors. For further references we present in table 3.4 the cases for which there are Majorana (M), pseudo-Majorana (M′), and Majorana–Weyl ($\bowtie$) spinors for $d \leqslant 12$ and $q = 0, 1, 2$ time-like dimensions (complex Weyl spinors occur in all even dimensions and have thus not been tabulated).

The reader should have no trouble figuring out any other case according to his needs, using the results M, M′, W and $\bowtie$ above. Here we still call attention to two important $\bowtie$ cases:

$d = 10$, $q = 1$ which is responsible for ten-dimensional supersymmetric Yang–Mills theory which when reduced to four-dimensions produces a famous ultraviolet finite quantum field theory;

Table 3.4

q \ d	1	2	3	4	5	6	7	8	9	10	11	12
0	M	M				M′	M′	$\bowtie$	M	M		
1	M′	$\bowtie$	M	M				M′	M′	$\bowtie$	M	M
2		M′	M′	$\bowtie$	M	M				M′	M′	$\bowtie$

$d = 12, q = 2$, which in spite of causality problems (closed curves in the plane of the two time-like dimensions are both time-like and closed) has been considered in connection with maximal supergravity (see chapter 26).

At a more mundane level, the existence of Majorana spinors in $d = 4$, $q = 1$ is essential for $N = 1$ supergravity in our world. With this spinor-classification problem fully solved, we are now in a position to find the physically relevant supersymmetries.

4

Physical supersymmetries

In chapter 1 we have seen that the spin–statistics connection and the 'no-go' theorems restrict the physically possible supersymmetries. Specifically, a physically admissible supersymmetry $\mathscr{s}$ must satisfy the following two principles:

(I) The Bose sector ${}^0\mathscr{s}$ of $\mathscr{s}$ must be the direct sum $\mathscr{P} + \mathscr{g}$ of the Poincaré algebra $\mathscr{P}$ and of an 'internal' symmetry algebra $\mathscr{g}$.

(II) All elements of the Fermi sector ${}^1\mathscr{s}$ of $\mathscr{s}$ must transform like Lorentz spinors.

The Poincaré algebra itself is not simple, not even semisimple, so it might appear somewhat far-fetched to expect the simple superalgebras described in chapter 2 to play a central role in physics. Yet these algebras make a crucial appearance, and in fact fully determine all superalgebras relevant for physics. For that matter, a similar statement holds for the Poincaré algebra $\mathscr{P}$. The Poincaré group enters physics as the group of isometries of Minkowski space. If space–time were not perfectly Minkowski but de-Sitter or anti-de-Sitter instead, its curvature would not vanish, rather it would be constant. The isometry group however would become a simple group: $O(4,1)$ for de-Sitter, $O(3,2)$ for anti-de-Sitter space.[†] In the limit in which the constant curvature of the de-Sitter or anti-de-Sitter space goes to zero (so that its 'curvature radius' becomes infinite) either of these spaces tends to Minkowski space and in the limit the simple isometry group contracts to the Poincaré group. This Wigner–Inönü contraction (Inönü & Wigner 1953, Gilmore 1974) can be tracked down at the level of Lie algebras. Since similar constructions become possible for superalgebras, we recall this ordinary de-Sitter → Poincaré contraction.

Independently of whether one considers the de-Sitter or anti-de-Sitter case, one can decompose the corresponding Lie algebra into a Lorentz ($\mathscr{o}(3,1)$) algebra and a remainder (a part corresponding at the group level to the $O(3,2)/O(3,1)$ or $O(4,1)/O(3,1)$ coset space, or at the algebra level to a Lie triple system, Bishop & Crittenden 1964). If, as usually, one labels

[†] We assume here a space–time with one time and three space dimensions.

a basis of $o(3,2)$ or $o(4,1)$ as $M_{\mu\nu} = -M_{\nu\mu}$ $(\mu, \nu = 1,\ldots,5)$, this decomposition corresponds to the two sets of generators $M = (M_{12}, M_{13}, M_{14}, M_{23}, M_{24}, M_{34})$, $P = (M_{15}, M_{25}, M_{35}, M_{45})$. Symbolically, the Lie bracketing relations can be written as

$$[M,M] \sim M, \quad [M,P] \sim P, \quad [P,P] \sim M, \tag{4.1}$$

where, for example, by $[M,P] \sim P$ we mean that the bracket of any element from the set M with any element of the set P yields an element of the set P, etc.... The contraction now amounts to rescaling the generators in M and P according to the rule

$$M \to \bar{M} = M, \quad P \to \bar{P} = \lambda P, \tag{4.2}$$

so that we obtain (still symbolically),

$$[\bar{M},\bar{M}] \sim \bar{M}, \quad [\bar{M},\bar{P}] \sim \bar{P}, \quad [\bar{P},\bar{P}] \sim \lambda^2 \bar{M} \tag{4.3}$$

and in the limit $\lambda \to 0$ (curvature radius $\to \infty$) we find precisely the commutation relations of the Poincaré algebra: momenta ($= P$s) commute.

Now we proceed to generalize this construction to the supersymmetric case (Freund & Kaplansky 1976). Naively one would search for de-Sitter superalgebras among the simple superalgebras that have as Bose sector the direct sum of the appropriate real form ($o(3, 2)$) or ($o(4, 1)$) of $o(5)$ and of an internal symmetry algebra. Even before confronting the existence problem for the requisite real form, we note a fatal flaw in this approach. The only simple superalgebras that qualify are of the form $osp(5|N)$ which in the Fermi sector contains N Lorentz vectors and N Lorentz scalars, thus violating principle II enunciated at the beginning of this chapter. We are led to the discouraging conclusion that our naive construction fails. What saves the day is an 'accident'. We are in a sufficiently low dimension to avail ourselves of the 'accidental' isomorphisms of low-dimensional Lie algebras. Indeed, the anti-de-Sitter Lie algebra $o(3,2)$ (but not the de-Sitter algebra $o(4,1)$!) is 'accidentally' isomorphic to the non-compact symplectic algebra $sp(4,R)$. So in addition to the unsuccessful choice $osp(5|N)$ above, this allows the extra choice $osp(N|4)$, and as we shall presently check, this choice conforms (after contraction) to both principles I and II.

Just as in the de-Sitter $\to$ Poincaré contraction described above, we start by dividing the generators of the extended super-de-Sitter algebra $osp(N|4)$ into four classes. In the Bose sector there are the (anti-)de-Sitter generators M and P of $sp(4,R) \sim o(3,2)$ and the generators B of the internal $o(N)$ symmetry. In addition there are, of course, the Fermi generators, the Qs. In the same symbolic notation as for the ordinary case, we can write down the $\mathcal{osp}(N|4)$ bracketing relations (as dictated by the construction in chapter 2) as

$$\begin{array}{lll} [M,M]\sim M & [B,M]\sim 0 & [M,Q]\sim Q \\ [M,P]\sim P & [B,P]\sim 0 & [P,Q]\sim Q \\ [P,P]\sim M & [B,B]\sim B & [B,Q]\sim Q \\ & [Q,Q]\sim M+P+B & \end{array}$$

where by $M+P+B$ we mean appropriate linear combinations of generators from the classes M, P, B. The contraction rescalings are now

$$\bar{M}=M, \quad \bar{P}=\lambda P, \quad \bar{Q}=\lambda^q Q, \quad \bar{B}=\lambda^b B.$$

For the $[Q,Q]$ bracket to have a nonsingular contracted ($\lambda\to 0$) limit we must impose $q\geqslant\frac{1}{2}$ and $b\leqslant 2q$. Regularity of the contracted $[B,B]$ and $[B,Q]$ brackets further requires $b\geqslant 0$. For $q>\frac{1}{2}$ the contractions are rather trivial: for $b<2q$ $[Q,Q]=0$, and for $b=2q$ $[Q,Q]=B$, but all Bs are central charges (they commute with all elements of the superalgebra). For $q=\frac{1}{2}$ and $0<b<2q=1$ the contracted superalgebra becomes a direct sum of an ordinary abelian algebra of Bs and a superalgebra of M, P, Q. The only interesting cases are thus

$$q=\tfrac{1}{2}, \quad b=0$$

and

$$q=\tfrac{1}{2} \quad b=1.$$

Let us first consider the case $b=0$. Then, in the limit $\lambda\to 0$ we obtain (we skip the bars over the symbols)

$$\begin{array}{lll} [M,M]\sim M & [M,B]\sim 0 & [M,Q]\sim Q \\ [M,P]\sim P & [P,B]\sim 0 & [B,Q]\sim Q \\ [P,P]\sim 0 & [B,B]\sim B & [P,Q]\sim 0 \\ & [Q,Q]\sim P & \end{array}$$

Notice that the M, P, Q sets span a closed superalgebra: the N-fold extended Poincaré superalgebra, and that the $\mathcal{o}(N)$ algebra (generators: B) enters only as an afterthought, as it were: it is represented on the Fermi sector of the superalgebra. It is also clear that one can truncate $\mathcal{o}(N)$ to any of its subalgebras without doing any harm. So after contraction the Bs need not span $\mathcal{o}(N)$, but rather any compact ordinary algebra unitarily representable on the Fermi sector.

Now to the alternative $b=1$. One gets the same results as for $b=0$

except that now

$$[B,B] \sim 0 \quad [B,Q] \sim 0 \quad [Q,Q] \sim P + B$$

This time around, the Bs commute with each other and with the whole rest of the superalgebra: they are central charges, whereas the $[Q,Q]$ bracket indicates a central extension of the N-fold extended Poincaré superalgebra. One can, of course, have both central charges and internal symmetries represented on the Fermi sector, by contracting part of the B sector with $b=0$ and part with $b=1$. Independently of this construction via the extended de-Sitter superalgebras, it can be shown that the N-fold extended Poincaré superalgebra with central charges and internal symmetries represented on the Fermi sector is the most general supersymmetry of a relativistic S-matrix theory with finite multiplets of massive particles (Hagg, Łopuszanski & Sohnius 1975). Any other type of supersymmetry would lead to problems such as those discussed in chapter 1. This is not surprising in view of the principles I and II we used in the construction. It should be noted here that in the conformally invariant case (when all masses vanish), the supersymmetry can be increased to $\mathfrak{su}(2,2|N)$, as we shall see below.

Here we first fill in some of the details. Of the bracketing relations presented above, those of the Poincaré algebra are well known. Then the commutators of the Lorentz generators (the Ms) with the Fermi generators (Qs) instruct us that the latter transform like spinors. Each Q has four spinor components and there are N such four-component spinors. These spinors are real, i.e., Majorana (this is most readily seen by noticing that before the contraction it is $\mathfrak{sp}(4,\mathbb{R})$ that enters $\mathfrak{osp}(N|4)$, so that the spinors are real.) So let Q^i_α ($\alpha = 1,\ldots,4$; $i = 1,\ldots,N$) represent the $4N$ components of these N Majorana spinors. The nontrivial new bracket is then

$$[Q^i_\alpha, Q^j_\beta] = 2(\gamma^\mu C)_{\alpha\beta}\delta^{ij}P_\mu + C_{\alpha\beta}Z^{[ij]} + (\gamma_5 C)_{\alpha\beta}Y^{[ij]} \tag{4.4}$$

where $Z^{[ij]}$ and $Y^{[ij]}$ are central charges. $[Q^i_\alpha, Q^j_\beta]$ is an anticommutator and as such the left-hand side is symmetric under the simultaneous interchanges $\alpha \leftrightarrow \beta$ and $i \leftrightarrow j$. This symmetry is readily confirmed also on the right-hand side keeping in mind that $(\gamma^\mu C)^T = \gamma^\mu C$, $C^T = -C$, $(\gamma_5 C)^T = -\gamma_5 C$, and that by definition, $Z^{[ij]} = -Z^{[ji]}$, $Y^{[ij]} = -Y^{[ji]}$. From these bracketing relations we see that (in the absence of central charges) the generators P^μ of translations are quadratic in the fermionic generators, which thus act as *de facto* 'square roots' of translations and are often referred to as *supertranslations.*

We now briefly consider the conformal superalgebras in four space–time dimensions. The ordinary conformal algebra in four dimensions is $\mathscr{o}(4,2)$ and again a naive orthosymplectic embedding violates principle II. This time again, we have the 'accidental' Lie algebra isomorphism $\mathscr{so}(4,2) \sim \mathscr{su}(2,2)$, which allows the embedding into $\mathscr{su}(2,2|N)$ ($\mathscr{P}\mathscr{su}(2,2|4)$ for $N = 4$), which turns out to be alright. One has the Lorentz generators M, the translations P, the conformal generators K, and the dilatation D, the supertranslations Q, the superconformal transformations R (the 'square roots' of the Ks), the internal $\mathscr{su}(N)$ transformations B, and the internal $\mathscr{u}(1)$ transformation C. Schematically, the brackets are:

$$
\begin{array}{llllllll}
[M,M]\sim M \\
[M,P]\sim P & [P,P]\sim 0 \\
[M,K]\sim K & [P,K]\sim M+D & [K,K]\sim 0 \\
[M,D]\sim 0 & [P,D]\sim P & [K,D]\sim K & [D,D]\sim 0 \\
[M,B]\sim 0 & [P,B]\sim 0 & [K,B]\sim 0 & [D,B]\sim 0 & [B,B]\sim B \\
[M,C]\sim 0 & [P,C]\sim 0 & [K,C]\sim 0 & [D,C]\sim 0 & [B,C]\sim 0 & [C,C]\sim 0 \\
[M,Q]\sim Q & [P,Q]\sim 0 & [K,Q]\sim R & [D,Q]\sim Q & [B,Q]\sim Q & [C,Q]\sim Q & [Q,Q]\sim P \\
[M,R]\sim R & [P,R]\sim Q & [K,R]\sim 0 & [D,R]\sim R & [B,R]\sim R & [C,R]\sim R & [Q,R]\sim M+D+B+C & [R,R]\sim K
\end{array}
$$

Clearly, the Poincaré superalgebra can be constructed in dimensions other than four as well (Nahm 1978), although in general, simple de-Sitter and/or conformal superalgebras will *not* be available. After all, we had to use accidental isomorphisms between ordinary Lie algebras, and their number is very limited.

Historical note

As mentioned above, Lie superalgebras have appeared, though not in a central role, in some mathematical contexts in the sixties (Fröhlicher & Nijenhuis 1956, Gerstenhaber 1963, 1964, Milnor & Moore 1965, Gell-Mann & Ne'eman 1964, Michel 1969). They were independently rediscovered in physics, and only following this, did the extensive physical and mathematical investigations, that form the subject of this book, kick off. Strictly speaking the first appearance of Lie superalgebras in a central role in a physical model is in the work of H. Miyazawa (1968). It is remarkable that in the context of an approximate unified model of mesons, baryons, antibaryons and exotic ($qq\bar{q}\bar{q}$) mesons Miyazawa gave a precise definition of a Lie superalgebra including the technical observation concerning the superselection rule mentioned in chapter 2. He then constructed the Lie algebra $V(6,21)$ which today would be called $\mathscr{su}(6|21)$. Work by Stavraky (1966) also dates back to this early epoch. Though unknown in the physics community, it seems to have triggered the mathematical investigations of V.G. Kac (1975), (1977). Next we have the inspired work of Gol'fand & Likhtman (1971), who discovered the four-dimensional Poincaré superalgebra. In spite of its remarkable results this paper was also ignored for a few years. Supersymmetry finally made it into the mainstream of physics in the heat of the activity in string theory that started the seventies. Ramond (1971) and Neveu & Schwarz (1971) discovered superstrings and the corresponding supersymmetric extensions of the infinite-dimensional Virasoro algebra (Virasoro 1970). Gervais & Sakita (1971) related this super-Virasoro algebra to invariances of the two-dimensional local field theory of the superstring. After the discovery of quantum chromodynamics (see for example, Marciano & Pagels 1978), with interest in hadronic strings on the wane, supersymmetry was being revived in a field-theoretic setting in a nonlinear realization by Volkov & Akulov (1973) and in the more direct linear realization by Wess & Zumino (1974). Remarkably neither of these pairs of authors were aware of the earlier work of Gol'fand & Likhtman. Like many another major concept in the sciences, supersymmetry has been discovered independently by various authors approaching different problems for seemingly different reasons. It was the Wess–Zumino work that opened the floodgates, thus setting the extremely fast pace for subsequent developments.

The next major technical step was taken by Salam & Strathdee (1974). They introduced superspace. At the mathematical end, all this inspired the classification of simple supersymmetries (Kac 1975, 1977, Freund & Kaplansky 1976, Nahm & Scheunert 1976, Kaplansky 1980) and the construction of a mathematically adequate superspace theory by Rogers (1980), based on earlier but physically less directly useful work of Kostant (1977), Batchelor (1980) and de Witt (1984).

Arriving in an era in which gauging was already second nature to all practicing theoretical physicists, it may not be surprising that supergravity (i.e., gauged supersymmetry) arrived so soon after supersymmetry. Volkov & Akulov already mentioned both supergravity and the fermionic counterpart of the Brout–Englert–Higgs phenomenon. Yet $N = 1$ supergravity was first explicitly constructed (in second order formalism!) by Freedman, van Nieuwenhuizen & Ferrara (1976), and right thereafter recast in an elegant first order form by Deser and Zumino (1976). The nonrenormalization theorems (Wess & Zumino 1974*a*, Iliopoulos & Zumino 1974) have led to the construction of supersymmetric grand unified theories by Witten (1981), Dimopoulos & Raby (1981), Dine, Fischler & Srednicki (1981) and by Sakai (1981).

Finite supersymmetric quantum field theories came about through the work of many people: Gliozzi, Olive & Scherk (1977), Brink, Scherk & Schwarz (1977), Jones (1977), Poggio & Pendleton (1977), Gell-Mann & Schwarz (1977), Grisaru, Roček & Siegel (1980), Avdeev, Tarasov & Vladimirov (1980), Caswell & Zanon (1981), Alvarez-Gaumé & Freedman (1981), Mandelstam (1983), Howe, Stelle & West (1983), Brink, Lindgren and Nilsson (1983).

At the supergravity end we encounter $N=1$ conformal supergravity (Kaku, Townsend & van Nieuwenhuizen (1977)), superspace formalisms (Wess & Zumino 1977, Brink, Gell-Mann, Ramond & Schwarz 1978, Siegel 1978, 1979, Ogievetsky & Sokatchev 1978) and extended supergravities, culminating in the construction of the $N=8$ supergravities by Cremmer & Julia (1978) and by de Wit & Nicolai (1982). All this then coalesced with modern Kaluza–Klein theory (Appelquist Chodos & Freund (1985)) via eleven-dimensional $N=1$ supergravity (Cremmer, Julia & Scherk 1978, Freund & Rubin 1980). Add to this the return of superstrings (Green & Schwarz 1984, 1984*a*) and we have updated this long history to where we now seem to be standing. Lest the reader be misled into believing that he has just read a full history of supersymmetry, let me warn him that this note is incomplete, though hopefully not capriciously so.

5

Particle contents of supermultiplets

In the absence of central charges, the Fermi–Fermi bracketing relation of the Poincaré superalgebra for four-dimensional Minkowski space–time is (see equation (4.1))

$$[Q^i_\alpha, Q^j_\beta] = 2\delta_{ij}(\gamma^\mu C)_{\alpha\beta}P_\mu \quad \alpha, \beta = 1, \ldots, 4 \quad i, j = 1, \ldots, N.$$

To find the particle contents of the corresponding supermultiplets (Salam & Strathdee 1974*a*, 1975, Fayet & Ferrara 1977), consider first the case with mass $m \neq 0$. Then we can go to the particles' rest frame where $P_\mu = (-m, 0, 0, 0)$. Rescaling Q^i_α into $\tilde{Q}^i_\alpha = m^{-1/2}Q^i_\alpha$, we then find in the rest frame

$$[\tilde{Q}^i_\alpha, \tilde{Q}^j_\beta] = 2\delta_{ij}\delta_{\alpha\beta} \equiv 2\delta_{i\alpha, j\beta} \tag{5.1}$$

which are precisely the anticommutation laws of the Clifford algebra $C(4n, 0)$. Its unique irreducible representation is 2^{2N}-dimensional. This 2^{2N}-dimensional supermultiplet contains both bosons and fermions and we now wish to find the corresponding spin assignments. At this point it is useful to switch from the Majorana presentation used until now, to an – in four dimensions, the case under consideration – equivalent Weyl presentation. This equivalence essentially expresses the fact that four real numbers can be replaced by two complex numbers.

Consider the Majorana (i.e., real) representation of $C(3, 1)$

$$\gamma^0 = \mathrm{i}\rho_2 \otimes \sigma_1 \quad \gamma^1 = \rho_1 \otimes \sigma_0 \quad \gamma^2 = \rho_2 \otimes \sigma_2 \quad \gamma^3 = \rho_3 \otimes \sigma_0$$

where ρ_i and σ_i are two copies of the usual basis for Pauli matrices and ρ_0 and σ_0, are two copies of the 2×2 unit matrix. Then

$$\gamma^5 = \gamma^0\gamma^1\gamma^2\gamma^3 = -\mathrm{i}\rho_2 \otimes \sigma_3.$$

Now, from any Majorana spinor

$$M_\alpha = M^*_\alpha$$

we construct the spinor

$$\mathrm{W} = \tfrac{1}{2}(1 - \mathrm{i}\gamma_5)\mathrm{M} \tag{5.2a}$$

with the result

$$\begin{pmatrix} W_1 \\ W_2 \\ W_3 \\ W_4 \end{pmatrix} = \tfrac{1}{2} \begin{pmatrix} M_1 + iM_3 \\ M_2 - iM_4 \\ i(M_1 + iM_3) \\ -i(M_2 - iM_4) \end{pmatrix}. \tag{5.2b}$$

Only two of the complex components of W, say W_1 and W_2 are independent and we assemble them into the two-component Weyl spinor

$$\begin{pmatrix} W_1 \\ W_2 \end{pmatrix} = \tfrac{1}{2} \begin{pmatrix} M_1 + iM_3 \\ M_2 - iM_4 \end{pmatrix} \tag{5.3}$$

which we henceforth denote as W_α, $\alpha = 1, 2$. Using the complementary projection operator $\frac{1}{2}(1 + i\gamma_5)$ and performing a similar truncation, we obtain a two-component spinor

$$\bar{W}_{\dot{\alpha}} = (W_\alpha)^* \quad \alpha = 1, 2, \tag{5.4}$$

whose components are the complex conjugates of those of W. Equations (5.2), (5.3) permit us also to recover a Majorana spinor from the equivalent Weyl spinor. In terms of W^i, and $\bar{W}^i$ the Weyl equivalents of the rest frame Majorana charges Q^i, the bracketing relations (5.1) become

$$[W^i_\alpha, W^j_\beta] = [\bar{W}^i_{\dot{\alpha}}, \bar{W}^j_{\dot{\beta}}] = 0$$
$$[W^i_\alpha, \bar{W}^j_{\dot{\beta}}] = \delta_{ij}\delta_{\alpha\beta}. \tag{5.5}$$

These are precisely the anticommutation relations of $2N$ pairs of Fermi creation–annihilation operators. The possible states can then all be obtained from a 'vacuum' state $|0\rangle$ on which the generators of the Clifford algebra (5.5) act. This Clifford vacuum is defined by

$$W^i_\alpha|0\rangle = 0 \quad \alpha = 1, 2 \quad i = 1, 2, \ldots, N$$

The possible states are then

$$|n_{11}n_{12}\ldots n_{2N}\rangle = \prod_{\substack{\alpha=1,2 \\ i=1,2,\ldots,N}} (\bar{W}^i_{\dot{\alpha}})^{n_{\alpha i}}|0\rangle$$

where on account of (5.5), each $n_{\alpha i}$ can take only two values: 0 and 1. There are thus a total of 2^{2N} states. These states can readily be classified by first noticing that the $4N^2$ operators $\bar{W}^i_{\dot{\alpha}} W^j_\beta$ generate under commutation the Lie algebra $U(2N)$. With respect to its $SU(2N)$ part, the 2^{2N} Clifford states are in a reducible representation: the direct sum of all totally antisymmetric representations:

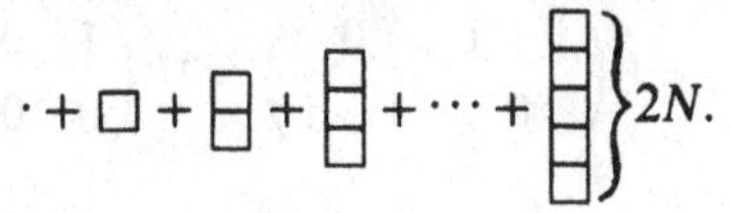

$$\cdot + \Box + \begin{array}{c}\Box\\\Box\end{array} + \begin{array}{c}\Box\\\Box\\\Box\end{array} + \cdots + \left.\begin{array}{c}\Box\\\Box\\\Box\\\Box\\\Box\end{array}\right\} 2N.$$

To obtain the spin content, consider that $SU(2)_{\text{spin}} \times SU(N)_{\text{internal}}$ subgroup of $SU(2N)$ for which the W^i are in the doublet (spin one-half) representation of $SU(2)$ and fundamental **N** representation of $SU(N)$. The branching rules of $SU(2N)$ multiplets into this $SU(2) \times SU(N)$ then determine the spin content, as can be verified by using the super-Poincaré commutation relations of the rotation generators with the supertranslation generators $\bar{W}^i_{\dot{\alpha}}$. Thus, e.g., for $N = 1$

$$SU(2N) = SU(2) \rightarrow SU(2) \times SU(N) = SU(2) \times SU(1) = SU(2)$$

and the $2^{2N} = 4$ states are in the representations

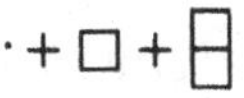

$$\begin{array}{ccccc} \cdot & + & \Box & + & \begin{array}{c}\Box\\\Box\end{array} \\ 0 & & \frac{1}{2} & & 0 \end{array} \quad \text{of } SU(2),$$

one spin one-half and two spin zero states. For $N = 2$

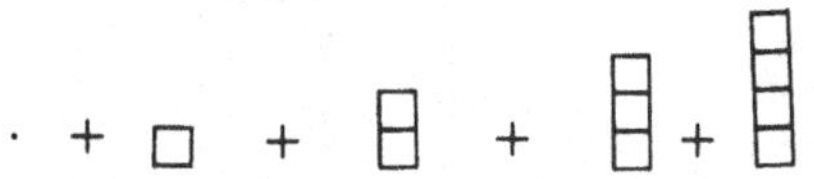

$$\begin{array}{lccccc} & \cdot \;+ & \Box \;+ & \begin{array}{c}\Box\\\Box\end{array} \;+ & \begin{array}{c}\Box\\\Box\\\Box\end{array} + & \begin{array}{c}\Box\\\Box\\\Box\\\Box\end{array} \\ SU(2N) = SU(4) & 1 & 4 & 6 & 4 & 1 \\ \downarrow & & & & & \\ SU(2) \times SU(2) & (0,0) & (\frac{1}{2},\frac{1}{2}) & (1,0)+(0,1) & (\frac{1}{2},\frac{1}{2}) & (0,0) \end{array}$$

corresponding to one spin one, four spin one-half and five spin zero states. Higher N can now be analyzed along identical lines. A discrete space inversion can also be defined and parities assigned.

We now discuss the massless supermultiplets. For $m = 0$ there is no rest frame and it becomes convenient to go to a frame in which the four-momentum P^μ is of the form

$$P^\mu = (|P|, 0, 0, P)$$

so that

$$P_\nu = (-|P|, 0, 0, P).$$

In Weyl notation, the only nontrivial Fermi–Fermi bracketing relations are

$$[W^i_\alpha, \bar{W}^j_{\dot{\beta}}] = 2\delta^{ij}\sigma^\mu_{\alpha\dot{\beta}}P_\mu$$

where $-\sigma^0$ is the unit 2×2 matrix and $\sigma^1, \sigma^2, \sigma^3$ are the Pauli matrices, so that (for $P>0$)

$$\sigma^{\mu}P_{\mu}=P\begin{pmatrix}1+1 & 0\\ 0 & 1-1\end{pmatrix}=2P\begin{pmatrix}1 & 0\\ 0 & 0\end{pmatrix},$$

and then

$$[W^{i}_{\alpha}, \bar{W}^{j}_{\beta}]=4P\delta^{ij}\delta_{\alpha 1}\delta_{\beta 1}.$$

After the renormalization $W^{i}_{\alpha}\to W^{i}_{\alpha}/(2P)^{1/2}$, $\bar{W}^{i}_{\beta}\to W^{i}_{\beta}/(2P)^{1/2}$, the only nontrivial brackets are

$$[W^{i}_{1}, \bar{W}^{j}_{1}]=\delta^{ij},$$

so there are only 2^{N} states (the $\bar{W}^{i}_{2}$ generate zero norm states). In this massless case the states are classified according to helicity rather than spin. One introduces again a 'Clifford vacuum' and the actual helicity assignments will depend on the helicity of this vacuum state. We present in table form the helicity assignments for $N=1,2,\dots,8$.

Table 5.1 *Number of states of given helicity for $N = 1, 2,\dots,8$*

helicity \ N	1	2	3	4	5	6	7	8
$j^{\max}$	1	1	1	1	1	1	1	1
$j^{\max}-\frac{1}{2}$	1	2	3	4	5	6	7	8
$j^{\max}-1$		1	3	6	10	15	21	28
$j^{\max}-\frac{3}{2}$			1	4	10	20	35	56
$j^{\max}-2$				1	5	15	35	70
$j^{\max}-\frac{5}{2}$					1	6	21	56
$j^{\max}-3$						1	7	28
$j^{\max}-\frac{7}{2}$							1	8
$j^{\max}-4$								1

These assignments, obtained from the representation theory of Clifford algebras, do not, in general, respect the CPT theorem, and as such 'doubled' reducible representations have to be considered to conform to this fundamental theorem of local relativistic quantum field theory. Such doubling can be avoided only for even N and $j^{\max}=\frac{1}{4}N$. These CPT self-conjugate supermultiplets for $N=2$, 4, 8 play a very special role, as we shall see.

The Clifford vacuum itself can be assigned attributes other than its helicity. It can transform according to a nontrivial representation of the internal symmetry, with the obvious changes in the counting of states.

From their construction, in both the massive and massless cases, the

Fermi and Bose states of a Poincaré supermultiplet are in correspondence with the odd and even sectors of a Clifford algebra respectively. These sectors have the same dimensionality (one-half the dimensionality of the full Clifford algebra). Hence in a supersymmetric theory the numbers of Bose and Fermi states in any supermultiplet must be the same. This strict balance between Bose and Fermi states (or degrees of freedom) will be repeatedly used in what follows.

6
Superspace

To come by a method of constructing manifestly supersymmetric action principles, it becomes advisable to generalize the concept of Minkowski space – or more generally, of Lorentzian manifold – in a way allowing for a simple action of the Poincaré, or of some other, supergroup. This generalization runs under the name of *superspace* (Salam & Strathdee 1974).

There exists a considerable body of work aiming to place superspace techniques (including supergroups, supermanifolds, superfields,...) on a mathematically solid footing (Kostant 1977, Batchelor 1980, de Witt 1984, Dell & Smolin 1979, Rogers 1980, 1981, Hoyos, Quirós, Ramirez Mittelbrunn & de Urriés 1984, Berezin 1979). We shall follow here and in chapters 7, 8 the work of Rogers (1980, 1981), which incorporates the features needed in physics.

In chapter 3 we have considered the Clifford algebra $C(Q_L)$ for the quadratic form Q_L. In the case $Q_L = 0$, $C(Q_L)$ becomes a Grassmann algebra B_L. Such a Grassmann algebra has L generators v_i, $i = 1, 2, \ldots, L$, obeying

$$v_i v_j + v_j v_i = 0. \tag{6.1}$$

In a basis of the form (3.2) an arbitrary element a of B_L is written as

$$a = a_0 \mathbf{1} + \sum_\Gamma a_\Gamma v_\Gamma$$

where

$$v_\Gamma \equiv v_{i_1 i_2 \ldots i_m} \equiv i^{m(m-1)/2} v_{i_1} v_{i_2} \cdots v_{i_m} \quad i_1 < i_2 < \cdots < i_m, \tag{6.2}$$

together with the scalars, form a 2^L-dimensional basis of B_L and the α_0, α_Γ are ordinary real numbers. The Grassmann algebra (like any Clifford algebra for that matter) admits a mod 2 grading:

$$B_L = {}^0B_L + {}^1B_L \tag{6.3}$$

with

${}^0B_L$ = set of all a for which $a_\Gamma = 0$ for all $v_\Gamma = v_{i_1 \ldots i_m}$ with m odd
${}^1B_L$ = set of all a for which $a_0 = 0$ and $a_\Gamma = 0$ for all $v_\Gamma = v_{i_1 \ldots i_m}$ with m even.

The elements of ${}^0B_L$ commute with all elements of B_L, whereas any two elements of ${}^1B_L$ anticommute. Both ${}^0B_L$ and ${}^1B_L$ are 2^{L-1}-dimensional. We now define the flat superspace $B_L^{4,4}$ corresponding to ordinary four-dimensional Minkowski space and allowing an action of the $N=1$ Poincaré superalgebra (four Fermi generators) by setting

$$\mathrm{B}_L^{4,4} = {}^0B_L \times {}^0B_L \times {}^0B_L \times {}^0B_L \times {}^1B_L \times {}^1B_L \times {}^1B_L \times {}^1B_L$$

where by $\times$ we mean the cartesian (not tensor) product. An arbitrary element or 'point' of $B_L^{4,4}$ is then given by specifying four even elements $x^0, x^1, x^2, x^3 \in {}^0B_L$ and for odd elements $\theta^1, \theta^2, \theta^3, \theta^4 \in {}^1B_L$. We emphasize that x^0, x^1, x^2, x^3 are general even elements of B_L, *not* ordinary real numbers as in special relativity. The dimension of $B_L^{4,4}$ is thus $8 \times 2^{L-1} = 2^{L+2}$. At this point the number L of Grassmann generators is as yet unspecified. Ultimately, we shall let $L \to \infty$ for reasons that will become clear below. As we considered a *real* B_L, the θ^α, $\alpha = 1,\ldots,4$ are real, so they are fit to span a Majorana spinor. It is convenient to use the notation z^M, $M = 1,\ldots,8$, for the coordinates of a superpoint:

$$x^0 \equiv z^4, x^1 \equiv z^1, x^2 \equiv z^2, x^3 \equiv z^3, \theta^\alpha \equiv z^{\alpha+4} \quad \alpha = 1,\ldots,4 \tag{6.4}$$

To qualify as a 'super-Minkowski space' $B_L^{4,4}$ must be endowed with a topology. On B_L itself we can define a norm. The norm $\|a\|$ of the element a, as given by the expansion (6.2) is defined as

$$\|a\| = |a_0| + \sum_\Gamma |a_\Gamma|.$$

It has the Banach (algebraic) properties

$$\|1\| = 1 \quad \|ab\| \leqslant \|a\|\,\|b\|$$

This norm induces a Hausdorff (separable) topology on B_L. In turn, this topology on B_L, induces a topology on $B_L^{4,4}$. Endowed with this topology, $B_L^{4,4}$ is now an adequate 'super' generalization of Minkowski space. More generally, we can define $B_L^{m,n}$ as the product of m copies of ${}^0B_L$ and of n copies of ${}^1B_L$. With n the dimensionality of spinors in m-dimensional Minkowski (or euclidean,...) space, $B_L^{m,n}$ will be the *flat* 'Minkowski superspace' in m-dimensions. In equations (6.4) $M = 1,\ldots,m+n$ with $z^1 \equiv x^1,\ldots,z^{m-1} \equiv x^{m-1}$, $z^m = x^0$, $z^{\alpha+m} \equiv \theta^\alpha$, $\alpha = 1,\ldots,n$. In what follows, we will stay with this more general case.

Once in possession of a superanalogue of flat Minkowski space it is natural to consider *supermanifolds* that look only locally like flat superspace. A (m,n)-dimensional supermanifold $\mathscr{M}$ over B_L is a Hausdorff topological space with a set of charts $\{U_\alpha, \psi_\alpha\}$ such that

(i) $\bigcup_\alpha U_\alpha = \mathcal{M}$

(ii) each ψ_α is a homeomorphism (i.e., a one-to-one map continuous in both directions) of U_α onto an open subset of $B_L^{m,n}$ under the just defined topology.

(iii) The transition functions $\psi_\alpha \circ \psi_\beta^{-1}: \psi_\beta(U_\alpha \cap U_\beta) \to \psi_\alpha(U_\alpha \cap U_\beta)$ are *superanalytic*.

By a superanalytic function $f: U \to B_L$, where U is an open set of $B_L^{m,n}$, we mean a function such that for any point p of coordinates z^M in U there exists a neighborhood V_p of p with the property that at any point $p' \in V_p$ of coordinates z'^M, $f(p')$ is given by the series expansion

$$f(p') = \sum_{k_1, \ldots, k_{m+n}=0}^{\infty} a_{k_1 \ldots k_{m+n}} (z^1 - z'^1)^{k_1} \ldots (z^{m+n} - z'^{m+n})^{k_{m+n}}$$

with $a_{k_1 \ldots k_{m+n}} \in B_L$. Obviously the requirement of superanalyticity can be suitably diluted, and more general classes of supermanifolds constructed. In particular the counterpart of C^∞ functions on ordinary manifolds are the G^∞ functions on supermanifolds. A function $f: U \to B_L$ is G^∞ on the open set $U \subset B_L^{m,n}$, if for the points of coordinates z^M and $z^M + a^M$ in U

$$f(z^M + a^M) = f(z^M) + \sum_{k=1}^{m+n} a^k G_k f(z^M) + O(\| a \|^2)$$

where the 'partial' superderivatives $G_K f: U \to B_L$ are themselves G^∞ functions. This is a recursive definition. Indeed, dropping the G^∞ requirements on the $G_k f$s we would obtain what may be called G^1 functions, imposing a G^1 property on the $G_K f$s defines G^2 functions, and so on. As examples, polynomials in z^M are G^∞ functions.

On an open subset V of a supermanifold, let $G^\infty(V)$ be the set of all G^∞ functions from V to B_L. A *vector field* X on V is now defined as a map $X: G^\infty(V) \to G^\infty(V)$ which is a superlinear superderivation, i.e.,

$$X(bg) = (-1)^{bx} b X g$$
$$X(fg) = (Xf)g + (-1)^{xf} f X g$$

for all f, $g \in G^\infty(V)$, $b \in B_L$. This of course is but the obvious super-generalization of the concept of a vector field on an ordinary manifold.

In the next chapter we utilize these definitions towards the construction of Lie supergroups, the 'exponentials' of Lie superalgebras with suitable Grassmann parameters.

7
Lie supergroups

An ordinary Lie group is obtained by exponentiating a Lie algebra with say real (or complex, etc.) parameters. In the case of superalgebras a similar exponentiation is contemplated but with parameters valued in a Grassmann algebra B_L, and with an implied pairing of even generators with even elements of B_L, and of odd generators with odd elements of B_L.

Technically, given a Lie superalgebra $\mathfrak{s}$ and a Grassmann algebra B_L we can consider extending the vector space of $\mathfrak{s}$ through left-multiplication by B_L. The Grassmann algebra B_L not being a field, but only a ring, the resulting object will not be a vector space but what in mathematics is called a module, or more accurately, due to the gradings, a *supermodule.* A left B_L-supermodule M is thus defined as an abelian group under addition, in which a distributive and associative left B_L-multiplication is defined. In other words to all $a \in B_L$ and $m \in M$ there exists $am \in M$ with the properties $a(m+n) = am + an$, $(a+b)m = am + bm$, $a(bm) = (ab)m$ for all $a, b \in B_L$ and $m, n \in M$. A Lie superalgebra W over R which is also a left B_L-supermodule, and for which the property $[ax, y] = a[x, y]$ for all $a \in B_L$, $x, y \in W$ holds, is called a *Lie supermodule* (or more precisely a left B_L Lie supermodule). The even part of such a Lie supermodule when exponentiated produces the *Lie supergroup.* Rather than go through this construction we shall directly define its endproduct and then retrace our way via left-invariant vector fields to this Lie supermodule. Thus we define a *Lie supergroup H* as:

- An abstract group which is also a superanalytic supermanifold of dimension (m, n) for which there exists a superanalytic mapping $H \times H \to H$ defined by

$$(h_1, h_2) \to h_1 h_2^{-1}.$$

For a given element $h \in H$ consider the map

- $L_h: H \to H$ via $L_h(k) = hk$ for all $k \in H$

L_h induces a map L_{h*} of vector fields on H, which to every vector field X

on H associates a new vector field $L_{h*}X$ according to the rule

$$(L_{h*}X)f \equiv X(f \circ L_h) \quad \text{for all} \quad f \in G^\infty(H).$$

A vector field X for which $L_{h*}X = X$ for all L_h is called *left-invariant*.

The set $\mathscr{L}(H)$ of all left-invariant vector fields on H is readily shown to span a Lie supermodule with the graded commutator of the vector fields as the bracket. As in the ordinary case $\mathscr{L}(H)$ is isomorphic to the tangent space to H at the unit element.

In the last section we observed that $B_L^{m,n}$ has ordinary dimension $(m+n)2^{L-1}$, so that a (m,n)-supermanifold can also be viewed as a $(m+n)2^{L-1}$-dimensional ordinary manifold. Superanalyticity then automatically implies ordinary analyticity. Thus a (m,n)-dimensional supergroup H over B_L can also be viewed as an ordinary $(m+n)2^{L-1}$-dimensional Lie group $\tilde{H}$. Now consider the $(m+n)2^{L-1}$-dimensional Lie algebra $\tilde{h}$ of $\tilde{H}$. On the other hand, note that in its turn the even part ${}^0\mathscr{L}(H)$ of the Lie supermodule $\mathscr{L}(H)$ can also be viewed as a $(m+n)2^{L-1}$-dimensional ordinary Lie algebra over $\mathbb{R}$. The theorem of Rogers (1981) states that

■ ${}^0\mathscr{L}(H)$ and $\tilde{h}$ are isomorphic.

We shall skip the proof and proceed instead to give a few examples.

Example 1: *The supergroups $GL(m|n,\mathbb{R})$ and $SL(m|n,\mathbb{R})$.* Consider the supergroup $GL(m|n,\mathbb{R})$ over B_L, the group of all nonsingular $(m+n)\times(m+n)$ matrices of the form

$$\begin{array}{c} \quad m \quad n \\ \begin{pmatrix} A & D \\ C & B \end{pmatrix} \begin{matrix} m \\ n \end{matrix} \end{array} \tag{7.1}$$

where all entries in the submatrices A, B are even elements of B_L and all entries of C and D are odd elements of B_L, and the group operation is ordinary matrix multiplication. It is straightforward to ascertain that $GL(m|n,\mathbb{R})$, so defined, is a Lie supergroup. In this case ${}^0\mathscr{L}(H)$ admits the basis $\varepsilon^{ij}G_{ij}$, where G_{ij} is the $(m+n)\times(m+n)$ matrix all entries of which vanish, except for the entry in the ith row and jth column, which equals one, and ε^{ij} is the most general even (odd) element of B_L if G_{ij} is an even (odd) element of $g\ell(m|n)$ as defined in chapter 2. Thus the $\varepsilon^{ij}G_{ij}$ span the ordinary $(m+n)^2 2^{L-1}$-dimensional Lie algebra ${}^0\mathscr{L}(GL(m|n,\mathbb{R}))$. Inspecting the defining matrices of $GL(m|n,\mathbb{R})$ we see that $g\ell(m|n,\mathbb{R})$ is the same Lie algebra.

For $m \neq n$ the supergroup $SL(m|n,\mathbb{R})$ is obtained from $GL(m|n,\mathbb{R})$ by imposing the condition that the '*superdeterminant*' (or Berezinian) of the matrix (7.1) be equal to one. The superdeterminant of a matrix M is defined as

$$\operatorname{s det} M = \exp(\operatorname{str} \ln M).$$

For a matrix M of the form (7.1) this yields

$$\operatorname{s det} M = \frac{\det A}{\det(B - CA^{-1}D)}.$$

Example 2: The (4, 4)-*dimensional translation supergroup* $T_{4,4}$. Consider $B_L^{4,4}$ itself with coordinates x^μ, θ^α and the group operation

$$(x^\mu, \theta^\alpha)(y^\mu, \varepsilon^\alpha) = (x^\mu + y^\mu + \mathrm{i}\bar{\theta}\gamma^\mu\varepsilon, \varepsilon^\alpha + \theta^\alpha). \tag{7.2}$$

Defining

$$\frac{\partial}{\partial\bar{\theta}^\alpha}\bar{\theta}^\beta = \delta_\alpha^\beta$$

the generators of ${}^0\mathscr{L}(T_{4,4})$ are

$$\varepsilon^\alpha Q_\alpha, \quad a^\mu P_\mu \tag{7.2a}$$

with $\varepsilon^\alpha (\alpha = 1, 2, 3, 4)$ four odd elements of $B_L^{4,4}$, a^μ $(\mu = 0, 1, 2, 3)$ four even elements of $B_L^{4,4}$ and

$$\left.\begin{aligned} Q_\alpha &= \frac{\partial}{\partial\theta^\alpha} + \mathrm{i}(\bar{\theta}\gamma^\mu C)_\alpha \frac{\partial}{\partial x^\mu} \\ P_\mu &= \mathrm{i}\frac{\partial}{\partial x^\mu}. \end{aligned}\right\} \tag{7.2b}$$

By taking the semidirect product of $T_{4,4}$ with the Lorentz group *Spin*(3, 1) the reader should have no difficulty reconstructing the (4, 4) Poincaré supergroups. The generators (7.2*b*) when supplemented by the Lorentz-algebra generators

$$M_{\mu\nu} = \mathrm{i}\left(x_\mu \frac{\partial}{\partial x^\nu} - x_\nu \frac{\partial}{\partial x^\mu}\right) + \bar{\theta}\sigma_{\mu\nu}\frac{\partial}{\partial\theta} \tag{7.2c}$$

provide a Schrödinger–Eckart representation of the (4, 4) Poincaré superalgebras.

8

Superfields

Just like ordinary fields are functions of the space–time coordinates x^μ, so superfields are functions of the superspace coordinates $z^M \equiv x^\mu, \theta^\alpha$. The θ^α being odd and therefore necessarily nilpotent elements of $B_L^{m,n}$, (the product of more than n θs is guaranteed to vanish), the superfields must be polynomials in them. The coefficients of the various θ-monomials in this θ-polynomial are functions only of the x^μs. This does *not* mean that they are ordinary numbers, the x^μs themselves being even elements of B_L!

We now proceed to make these ideas precise. To this end, given $a \in B_L$ define the *body* $\beta(a)$ of a as the coefficient a_0 of $\mathbf{1}$ in the expansion (6.2a) of a.

$$\beta(a_0\mathbf{1} + \sum_\Gamma a_\Gamma v_\Gamma) = a_0.$$

The *soul* $\sigma(a)$ of a is then defined as

$$\sigma(a) = a - \beta(a)\mathbf{1}.$$

In other words the body is the ordinary 'c-number' part of a, whereas the soul is the nilpotent part of a (de Witt 1984).

Consider now a point P in the superspace $B_L^{m,n}$ of coordinates $(x^0, \ldots, x^{m-1}, \theta^1, \ldots, \theta^n)$. The body $\beta(P)$ of P is defined as the point of $\mathbb{R}^m$ of coordinates $\beta(x^0), \ldots, \beta(x^{m-1})$, whereas $\sigma(P)$, the soul of P, is the point $(\sigma(x^0), \ldots, \sigma(x^{m-1}), \sigma(\theta^1), \ldots, \sigma(\theta^n))$ of $B_L^{m,n}$. For L finite but $L > n$, let U be an open set of $B_L^{m,n}$ and $\beta(U)$ the set of $\mathbb{R}^m$ corresponding to the bodies of the points of U. Let f be a C^∞ function from $\beta(U)$ to B_L. To any such function f we associate a function $z(f)$ from U itself to B_L, according to the rule

$$\begin{aligned} &z(f)[x^0, \ldots, x^{m-1}, \theta^1, \ldots, \theta^n] \\ &= \sum_{i_0=0}^{L} \cdots \sum_{i_{m-1}=0}^{L} \frac{1}{i_0! \ldots i_{m-1}!} \left[\left(\frac{\partial}{\partial \beta(x^0)} \right)^{i_0} \cdots \right. \\ &\left. \cdot \left(\frac{\partial}{\partial \beta(x^{m-1})} \right)^{i_{m-1}} f(\beta(x^0), \ldots, \beta(x^{m-1})) \right] (\sigma(x^0))^{i_0} \ldots (\sigma(x^{m-1}))^{i_{m-1}}. \end{aligned} \tag{8.1}$$

This $z(f)$ is a continuation (a G^∞ continuation at that, as is readily shown) of a function f over ordinary space into a function over superspace. Note though, that at this point the odd coordinates had not yet been called upon. The most general G^∞ function on U can now be built out of such z-continuations of C^∞ functions on ordinary space and of monomials in the odd θ^α coordinates. Specifically, all $F \in G^\infty(U)$ can be expanded in the form

$$F = \sum_{\Delta = 0, \alpha, \ldots} w_\Delta z(F_\Delta) \tag{8.2}$$

where $w_0 = 1$, $w_\alpha = \theta_\alpha$, $w_{\alpha_1 \alpha_2} = \theta_{\alpha_1}\theta_{\alpha_2}, \ldots, w_{\alpha_1, \ldots, \alpha_n} = \theta_{\alpha_1}\theta_{\alpha_2}\ldots\theta_{\alpha_n}$ and each F_Δ is a C^∞ function from the body $\beta(U)$ into B_L. The F_Δs are uniquely determined by F, uniquely that is, up to terms annihilated by the w_Δs. Conversely, every function admitting an expansion (8.2) is in $G^\infty(U)$ (Rogers 1980).

If a function ϕ maps U into the whole B_L but only into its even (odd) sector ${}^0B_L({}^1B_L)$, it is called an *even* (*odd*) *function*. An even (odd) function on flat superspace $B_L^{m,n}$ is called an *even* (*odd*) *superfield*. We shall deal almost exclusively with even superfields and shall refer to them simply as *superfields* (Salam & Strathdee 1974).

The superfields we are interested in must have simple supersymmetry transformation laws. Under an infinitesimal ordinary translation of parameter α^μ an ordinary scalar field transforms as

$$\delta\phi = \mathrm{i}\alpha^\mu \partial_\mu \phi.$$

Similarly, for a scalar superfield ϕ, a supersymmetry transformation of infinitesimal Fermi parameters ε^α (i.e., an element of ${}^0\mathscr{L}_{4,4}$) causes the change

$$\delta\phi = \mathrm{i}\varepsilon^\alpha Q_\alpha \phi$$

with Q_α as given by equations (7.2*b*). Similar formulae for ordinary translations and Lorentz transformations are also readily obtained. Of course one can consider superfields that are not scalars, but carry ordinary vector or spinor indices or superspace 'vector' indices.

More generally on a (4,4) supermanifold M, we have charts (U_α, ψ_α) to $B_L^{4,4}$, so that a superfield ϕ on U_α is defined by the superfield $\phi\psi_\alpha^{-1}$ on the open set $\psi_\alpha(U_\alpha)$ of $B_L^{4,4}$. Again an expansion (8.2) is always possible for a superfield.

We have finally come to the point where we can discuss the value L (the order of the Grassmann algebra B_L) to be used in physics. Consider a superfield ϕ. It admits an expansion (8.2), but being even, all the coefficients of w_Δs which are products of an odd number of θs are themselves odd elements of B_L, and as such necessarily nilpotent as long as L is finite. In the quantum theory these coefficients will correspond to Fermi fields, yet any product of more than L such fields will necessarily vanish, so that a truncation of Green's functions occurs which violates unitarity. The way out, is to let $L \to \infty$ thus eliminating this unwanted truncation. The technical details of the expansion (8.2) have to be slightly changed in the case $L = \infty$, one has to substitute entire functions for G^∞ and C^∞ functions. In fact, there are also mathematical reasons for taking the $L \to \infty$ limit (Hoyos, Quirós, Ramirez Mittelbrunn & de Urriés 1984).

9

Integration on Grassmann algebras

To formulate an action principle on superspace we have to integrate not over ordinary $\mathbb{R}$-valued coordinates, but over coordinates valued in some Grassmann algebra B_L. How is this to be done? Consider, for simplicity, one Fermi variable $\theta \in {}^1B_L$. The most general B_L-valued function of θ is of the form

$$f = a + b\theta$$

with $a, b \in B_L$, but independent of θ. The integral $\int f \, \mathrm{d}\theta$ is envisioned as generalizing the definite integral $\int_{-\infty}^{+\infty}$ over the ordinary coordinates, and as such required to be a map from the B_L-valued functions of θ to the real numbers, which is

(i) *linear*: $\int \sum C_i f_i(\theta) \mathrm{d}\theta = \sum C_i \int f_i(\theta) \mathrm{d}\theta$ with f_i functions of θ, $C_i \in B_L$ independent of θ, i.e., 'constants'.

(ii) *translationally invariant* $\int f(\theta + \varepsilon) \mathrm{d}\theta = \int f(\theta) \mathrm{d}\theta$ for all 'constant' $\varepsilon \in {}^1B_L$.

In detail this yields

$$\int (a + b\theta + b\varepsilon) \mathrm{d}\theta = (a + b\varepsilon) \int \mathrm{d}\theta + b \int \theta \mathrm{d}\theta = \int (a + b\theta) \mathrm{d}\theta = a \int \mathrm{d}\theta + b \int \theta \mathrm{d}\theta$$

where in the first and last steps we used linearity. This means

$$b\varepsilon \int \mathrm{d}\theta = 0, \quad b \int \theta \mathrm{d}\theta = \text{arbitrary}$$

leading to the definition (Matthews & Salam 1955, Candlin 1956, Berezin 1966)

$$\int \mathrm{d}\theta = 0 \quad \int \theta \mathrm{d}\theta = 1 \tag{9.1a}$$

The normalization of $\int \theta \mathrm{d}\theta$ is convenient but arbitrary; all that is required to avoid triviality is $\int \theta \mathrm{d}\theta \neq 0$. When more variables $\theta^1, \theta^2, \ldots, \theta^n$ are contemplated, (9.1a) implies

$$\int \theta^i \mathrm{d}\theta^j = \delta^{ij} \quad \int \mathrm{d}\theta^j = 0 \tag{9.1b}$$

Consider a function $f = a + b\theta$. We have then

$$\int f \mathrm{d}\theta = b$$

Taking the θ-derivative of f from the right

$$f\frac{\overleftarrow{\mathrm{d}}}{\mathrm{d}\theta} = b$$

we find the same result. Whereas ordinary indefinite integration is the inverse of derivation, this *Berezin integration* over θ is the *same* thing as derivation, but of course it generalizes definite, not indefinite integration. To find the connection of this integral with some sort of 'Riemann sum' is an interesting problem (Rabin 1984). We further note that in view of the normalization $\int \theta \mathrm{d}\theta = 1$, θ and $\mathrm{d}\theta$ have opposite dimensionalities. From the supersymmetry transformation law

$$\theta \to \theta + \varepsilon$$
$$x^\mu \to x^\mu + \mathrm{i}\bar{\theta}\gamma^\mu\varepsilon.$$

we see that θ and ε have the same dimensionality and that the coordinates x^μ have the square of that dimensionality. The body of the coordinate x^μ is the usual c-number coordinate and thus has dimension of length. Thus the fermionic coordinates of superspace, the θs, have dimension $L^{1/2}$ and finally $\mathrm{d}\theta$ dimension $L^{-1/2}$! As far as integration is concerned, each Bose coordinate x increases the dimension of superspace by one, each Fermi coordinate θ *decreases* it by one-half. It may then not be surprising that in supersymmetric quantum field theories in which integrals now sum over both Bose and Fermi dimensions convergence improves, in some cases all the way to yielding finite theories.

For future use we still define here a Grassmann variant of the Dirac 'δ-function' by

$$\int f(\theta)\delta(\theta)\mathrm{d}\theta = f(0) \tag{9.2}$$

so that

$$\delta(\theta) = \theta, \int \delta(\theta)\mathrm{d}\theta = 1 \quad \text{and} \quad \int \theta\delta(\theta)\mathrm{d}\theta = 0.$$

In case $n > 1$ Grassmann variables $\theta^1, \theta^2, \ldots, \theta^n$ are encountered, we define

$$\delta^n(\theta) = \delta(\theta^n)\delta(\theta^{n-1})\ldots\delta(\theta^1) = \theta^n\theta^{n-1}\ldots\theta^1$$

with all the obvious properties.

As noted in chapter 1, the action of a physical system is an integral over space–time. Similarly, in the superspace formulation of a supersymmetric field theory, the action is given as an integral over superspace. As we shall see, on account of the extreme simplicity of Berezin integration (equations (9.1)), the integrals over the n fermionic coordinates of superspace can be carried out explicitly. This then leaves an integral over the m bosonic coordinates. These coordinates, as we repeatedly emphasized, are *not* valued in the reals, rather they are even elements of B_L. Integrating over such ${}^0B_L$-valued coordinates, is also not an ordinary Riemann sum (de Witt 1984). It is most simply viewed as a contour integral in ${}^0B_L$, which however is contour independent, so that nothing is lost by integrating along the real valued bodies of the bosonic coordinates. This way we do have a full prescription for performing superspace integrals, one which is nicely consistent with changes of variables (e.g. from supersymmetry transformations) which affect the soul of the bosonic space–time coordinates. With the mathematical framework for the construction of supersymmetric theories thus in place, we proceed in the next chapter with the simplest case of supersymmetric point particles.

Part II

Globally supersymmetric theories

10

Supersymmetric point particle mechanics

The classical mechanics of point particles is so very simple, because it corresponds to a 'field theory' in one time and zero space dimensions. The particle's coordinates $x^1,\ldots,x^d$ are viewed as d 'scalar fields' of the one-dimensional time-variable t. To supersymmetrize this setup (Friedan & Windey 1984) all one need do is introduce a (1,1) superspace of coordinates (t,τ) with $t\in{}^0B_L$ and $\tau\in{}^1B_L$. The nonrelativistic superpoint particle is then described by d scalar superfields $X^1(t,\tau),\ldots,X^d(t,\tau)$. We have the expansion of type (8.2)

$$X^a(t,\tau)=x^a(t)+\theta^a(t)\tau \tag{10.1}$$

with $x^a\in{}^0B_L$ and $\theta^a\in{}^1B_L$. The supersymmetry generators are then

$$\begin{aligned} Q &= \mathrm{i}\tau\partial_t-\partial_\tau \\ H &= \mathrm{i}\partial_t \end{aligned} \tag{10.2}$$

so that the superalgebra is

$$[Q,Q]\equiv 2Q^2=-2H,\quad [Q,H]=[H,H]=0. \tag{10.3}$$

This is an algebra of left-supertranslations and time-translations. The corresponding right-supertranslations are generated by

$$D=\mathrm{i}\tau\partial_t+\partial_\tau. \tag{10.4}$$

Indeed

$$D^2=H,\quad [D,H]=[H,H]=0 \tag{10.5a}$$

but also

$$[Q,D]_+=0. \tag{10.5b}$$

On account of this last property, the generator D of right-translations can be utilized in the construction of a manifestly supersymmetric action:

$$S_1=\int_{t_i}^{t_f}\mathrm{d}t\int\mathrm{d}\tau\mathscr{L}_1 \tag{10.6a}$$

$$\mathscr{L}_1=\tfrac{1}{2}DX^aD(DX^a). \tag{10.6b}$$

That the supersymmetry of this action is manifest, is made clear by the following reasoning. Under a supersymmetry transformation of Fermi parameter α, the superfields X^a undergo the change

$$\delta X^a = \alpha Q X^a. \tag{10.7}$$

This can be used to calculate the change of the lagrangian superdensity $\mathscr{L}_1$. According to equation (10.5*b*) we have

$$[\alpha Q, D]_- = 0 \tag{10.8}$$

so that $\delta DX^a = \alpha QDX^a = D\alpha QX^a = D\delta X^a$, justifying the name covariant derivative for DX^a. From (10.7), (10.8) and (10.6*b*) we then find

$$\delta\mathscr{L}_1 = \alpha(\mathrm{i}\tau\partial_t - \partial_\tau)\mathscr{L}_1 \tag{10.9}$$

The term $\mathrm{i}\alpha\tau\partial_t\mathscr{L}_1 = \partial_t(\mathrm{i}\alpha\tau\mathscr{L}_1)$ is an exact time-derivative and as such does not contribute to variations. The term $-\alpha\partial_\tau\mathscr{L}_1$ on the other hand is τ independent, $\mathscr{L}_1$ being at most linear in τ. Its τ-integral then vanishes by Berezin's rule. Thus

$$\delta S_1 = \int \mathrm{d}t\mathrm{d}\tau\,\delta\mathscr{L}_1 = \text{'surface' term} \tag{10.10}$$

so that the action principle is supersymmetric.

It furthermore follows from (10.9) that $\mathscr{L}_1$, itself a product of two superfields, behaves under supersymmetry transformations as a superfield. This result is quite general, it only depends on the supersymmetry generator αQ acting as a derivation (obeying Leibniz' rule) on superfields. As such this argument holds in higher space–time dimensions as well: the product of two superfields is again a superfield.

To find the supersymmetry-transformation laws of the X^a and θ^a we return to formula (10.7) and using equations (10.1), (10.2) write

$$\delta X^a = \delta x^a + \delta\theta^a\tau = \alpha(\mathrm{i}\tau\partial_t - \partial_\tau)(x^a + \theta^a\tau) = \mathrm{i}\alpha\tau\dot{x}^a + \alpha\theta^a \tag{10.11a}$$

so that in component form

$$\delta x^a = \alpha\theta^a \quad \delta\theta^a = \mathrm{i}\alpha\dot{x}^a. \tag{10.11b}$$

From the definitions (10.1), (10.4), $DX^a = -\theta^a + \mathrm{i}\dot{x}^a\tau$, $DDX^a = \mathrm{i}(\dot{x}^a + \dot{\theta}^a\tau)$, so that the action S_1 itself can be put in component form, by explicitly carrying out the Berezin τ-integration:

$$S_1 = \tfrac{1}{2}\mathrm{i}\int \mathrm{d}t\int \mathrm{d}\tau(-\theta^a + \mathrm{i}\dot{x}^a\tau)(\dot{x}^a + \dot{\theta}^a\tau) = -\tfrac{1}{2}\int \mathrm{d}t\int \mathrm{d}\tau\,\tau(\dot{x}^a\dot{x}^a + \mathrm{i}\theta^a\dot{\theta}^a)$$

$$= \tfrac{1}{2}\int \mathrm{d}t(\dot{x}^a\dot{x}^a + \mathrm{i}\theta^a\dot{\theta}^a). \tag{10.12}$$

The first term is just the kinetic energy term of an ordinary point particle in which the mass m of the particle is $m=1$. The $\frac{1}{2}\theta^a\dot{\theta}^a$ term, dictated by supersymmetry, is new. It is a 'kinetic' piece corresponding to the superpoint particle's Grassmann degrees of freedom (the θ^as). Its appearance is somewhat unusual. If the θs were ordinary Bose variables, this would be a surface term. But for anticommuting quantities $2\theta^a\dot{\theta}^a = -2\dot{\theta}^a\theta^a \neq (\mathrm{d}/\mathrm{d}t)(\theta^a\theta^a) = 0$. The equations of motion obtained from varying the action (10.12) are second order for the Bose variables and first order for the Fermi variables.

$$\ddot{x}^a = 0 \quad \dot{\theta}^a = 0. \tag{10.13}$$

We are obviously dealing with a free superpoint particle. To include interaction we would want to generalize the ordinary potential term $V(x)$, say to $V(X(t,\tau))$. This doesn't work in this simple case since

$$V(X) = V(x) + \left.\frac{\partial V}{\partial X^a}\right|_{\tau=0} \cdot \theta^a \tau$$

and the term $V(x)$, whose supersymmetrization we seek, disappears upon τ-integration. This difficulty is connected with the absence of space dimensions. It can be corrected, by going to the $N=2$ case with two super-charges. To this end we switch to a $(1,2)$ superspace of coordinates (t,τ^1,τ^2). The superfields are now

$$X^a(t,\tau^\alpha) = x^a(t) + \theta^a_\alpha(t)\tau^\alpha + \mathrm{i}F^a(t)\tau^1\tau^2, \quad \alpha = 1,2;\ a = 1,\ldots,d. \tag{10.14}$$

The supersymmetry generators (see equations (7.2))

$$\left.\begin{aligned} Q_\alpha &= \mathrm{i}\tau^\alpha\partial_t - \partial_\alpha \quad (\partial_\alpha \equiv \partial/\partial\tau^\alpha) \\ H &= \mathrm{i}\partial_t \end{aligned}\right\} \tag{10.15}$$

now span the superalgebra

$$\left.\begin{aligned} [Q_\alpha, Q_\beta] &= -2\delta_{\alpha\beta}H \\ [Q_\alpha, H] &= [H,H] = 0. \end{aligned}\right\} \tag{10.16}$$

In terms of the right-supertranslations

$$D_\alpha = \mathrm{i}\tau^\alpha\partial_t + \partial_\alpha \tag{10.17}$$

the action of the free $N=2$ superpoint particle now becomes

$$S_2 = -\int \mathrm{d}t\,\mathrm{d}\tau^2\,\mathrm{d}\tau^1\,\tfrac{1}{4}\varepsilon_{\alpha\beta}D_\alpha X^a D_\beta X^a \tag{10.18}$$

which, after carrying out the τ-integrations, yields

$$S_2 = \int dt \tfrac{1}{2}(\dot{x}^a \dot{x}^a + i\theta^a_\alpha \dot{\theta}^a_\alpha + F^a F^a) \tag{10.19}$$

It is worth noting some differences between the actions S_1 and S_2. First of all, S_1 contains three covariant derivatives (one D on the first factor and DD on the second), whereas S_2 only contains two. This is readily accounted for by observing that S_1 contains only one Berezin integration while S_2 contains two. The action must be bosonic so the lagrangian density $\mathscr{L}_1$ must be fermionic and $\mathscr{L}_2$ bosonic. For $\mathscr{L}_1$, had we taken only one covariant derivative we would have obtained the 'surface' term $D(X^a X^a) = 2DX^a X^a = 2X^a DX^a$, so the term with three Ds is minimal. Moreover, D, D_α, $d\tau_1$ and $d\tau_2$ all have dimension $(\text{time})^{-1/2}$, so the correct dimension of action is only obtained for the number of derivatives that actually appear in S_1 and S_2. A further difference is the appearance in S_2 of the new bosonic variables F^1 and F^2. Their equations of motion, as obtained by varying S_2, are

$$F^1 = F^2 = 0, \tag{10.20}$$

so that F^1 and F^2 do not really evolve. They are *auxiliary* variables that can be eliminated.

For the action S_2, we can now add an interaction term on the sly, as it were. Adding to S_2 integral

$$S_{\text{int}} = i \int dt d\tau^2 d\tau^1 W[X^a(t, \tau^1, \tau^2)] \tag{10.21}$$

with W an arbitrary function of the superfields X^a, will maintain supersymmetry again by an argument of the type of equations (10.7)–(10.10). Naively one would expect $V(x(t))$ again to be swept away by the Berezin integrations. But the auxiliary fields come to the rescue. For, carrying out the τ^1 and τ^2 integrations we find

$$S_{\text{int}} = -\int dt \left\{ \frac{\partial W[x(t)]}{\partial x^a(t)} F^a - i \frac{\partial^2 W(y)}{\partial y^b \partial y^a}\bigg|_{y^a = x^a(t)} \cdot \theta^a_1 \theta^b_2 \right\} \tag{10.22}$$

so that the equations of motion for the auxiliary fields, while still algebraic rather than differential, are nevertheless less trivial now:

$$F^a = \frac{\partial W(x)}{\partial x^a}. \tag{10.23}$$

Inserting this in the action, we find

$$S \equiv S_2 + S_{\text{int}} = \int dt \left[\tfrac{1}{2}\dot{x}^a\dot{x}^a + \tfrac{1}{2}\mathrm{i}\theta^a_\alpha\dot{\theta}^a_\alpha - V(x^b) + \mathrm{i}\left.\frac{\partial^2 W(y)}{\partial y^b \partial y^a}\right|_{y^a = x^a(t)} \cdot \theta^a_1 \theta^b_2 \right] \tag{10.24a}$$

with the positive definite potential

$$V(x) = \frac{1}{2}\left.\frac{\partial W(y)}{\partial y^a}\frac{\partial W(y)}{\partial y^a}\right|_{y^b = x^b(t)}. \tag{10.24b}$$

We shall deal with this simple class of problems in some more detail in chapter 13. Here we content ourselves with this construction of our first supersymmetric action principle. We now go on to explore the physics of these somewhat unusual 'classical' systems with Grassmann algebra valued variables.

11

Pseudoclassical mechanics of superpoint particles

While Grassmann variables were forced upon us by considerations of supersymmetry, let us disregard supersymmetry for the moment, and analyze the 'classical' mechanics of systems with Grassmann degrees of freedom. Actually, this is not quite classical mechanics as we shall presently see, so it is usual to refer to it as 'pseudoclassical' or 'prequantum' mechanics (Berezin & Marinov 1977, Casalbuoni 1976, 1976*a*).

Let $\theta^1(t)$, $\theta^2(t)$, $\theta^3(t)$ be odd elements of the Grassmann algebra B_3. Consider an action involving only these Grassmann variables (no space coordinates)

$$\left.\begin{aligned} S &= \int \mathrm{d}t\,[\tfrac{1}{2}\mathrm{i}\theta^a\dot{\theta}^a - H(\theta^b)] \\ H(\theta^b) &= H(0) - \tfrac{1}{2}\mathrm{i}\varepsilon_{abc}\kappa B^a\theta^b\theta^c \end{aligned}\right\} \tag{11.1}$$

with $H(0)$, κ, $B^a(a = 1, 2, 3)$ all real numbers. The Euler–Lagrange equations are

$$\dot{\theta}^a = \mathrm{i}H\frac{\overleftarrow{\partial}}{\partial\theta^a} = \varepsilon_{abc}\kappa B^b\theta^c. \tag{11.2}$$

This is the well-known differential equation for the precession of an (admittedly Grassmann) vector in a magnetic field. It has the solution

$$\theta^a(t) = R^{ab}(t)\theta^b(0) \tag{11.3}$$

where $R^{ab}(t)$ is the element of 0(3) (the three-dimensional rotation group), corresponding to uniform rotation with precession frequency vector

$$\vec{\omega} = \kappa\vec{B}. \tag{11.4}$$

This all suggests that, textbook wisdom to the contrary, we are dealing here with the (pseudo)-*classical* mechanics of spin! That this is the correct interpretation will follow convincingly from the rest of our argument. It is convenient to define here a graded counterpart to Poisson's bracket

$$[f(\theta), g(\theta)]_{\mathrm{P}} = \mathrm{i}\left(f\frac{\overleftarrow{\partial}}{\partial\theta^a}\right)\left(\frac{\overrightarrow{\partial}}{\partial\theta^a}g\right). \tag{11.5}$$

It is graded antisymmetric

$$[f,g]_P = -(-1)^{fg}[g,f]_P \tag{11.6}$$

and obeys the graded Jacobi identity (of the type of equation (2.1)). Using $\dot{\theta}^a = iH(\overleftarrow{\partial}/\partial\theta^a)$ we find as usual

$$\dot{f} = [H,f]_P. \tag{11.7}$$

Finally the definition of $[\ \ ]_P$ entails

$$[\theta^a, \theta^b]_P = i\delta^{ab}. \tag{11.8}$$

The (spin) angular momentum components

$$S^a = -\tfrac{1}{2}i\varepsilon^{abc}\theta^b\theta^c \tag{11.9}$$

span, under Poisson bracketing, the $\mathscr{su}(2)$ Lie algebra

$$[S^a, S^b]_P = -\varepsilon^{abc}S^c \tag{11.10}$$

This spin vector $\vec{S}$ represents the total angular momentum of the system. Its projection $\vec{B}\cdot\vec{S}$ along the direction of the magnetic field is conserved, corresponding to the invariance of the system under rotations around the direction of $\vec{B}$. For the free spin ($\vec{B}=0$), all three components of $\vec{S}$ are conserved.

Space degrees of freedom $\vec{q}$ can be imparted to the particle, thus allowing for a more realistic action:

$$S = \int_{t_i}^{t_f} dt(\vec{p}\dot{\vec{q}} + \tfrac{1}{2}i\vec{\theta}\dot{\vec{\theta}} - \frac{\vec{p}^2}{2m} - V(\vec{q}) - \vec{L}\cdot\vec{S}V_{LS} - \kappa\vec{S}\cdot\vec{B}) \tag{11.11}$$

here $\vec{L} = \vec{q}\times\vec{p}$ is the orbital angular momentum, a spin–orbit potential V_{LS} and a central potential V have been introduced besides the magnetic field, and the particles's mass m has been explicitly exhibited. This action is simpler than it appears at first sight, the vector $\vec{S}$ being bilinear in the three θs, so that $\vec{S}^2$ and $(\vec{L}\,\vec{S})^2$, both contains four θs and as such *vanish*. As a consequence, the equations of motion can be simply solved.

We now address a question of principle. We were pretending so far, to be dealing with a classical system. Questions of measurement, which can be very subtle in quantum mechanics, are, as a rule, of no particular complexity in classical mechanics. Yet, here we talk about the time evolution of Grassmann algebra valued quantities. These are not ordinary numbers, whereas quantities measured in experiments are always valued in the reals. So we would like to associate to every Grassmann algebra valued observable $f(\theta)$ of our system, a real number: its 'expectation value'. To this end we introduce a system *density function* $\rho(\theta, t)$ obeying a Liouville equation

$$\frac{\partial \rho}{\partial t} + [H, \rho]_P = 0 \tag{11.12}$$

and use this density function to define the expectation value of $f(\theta)$ as

$$\langle f \rangle = \mathrm{i} \int f(\theta)\rho(\theta, t)\mathrm{d}^3\theta. \tag{11.13}$$

How do we choose ρ? Well, we require that only even elements of the Grassmann algebra have a nonvanishing expectation value, and that

$$\langle \mathbf{1} \rangle = 1, \quad \langle \vec{S} \rangle = \vec{C},$$

where $\vec{C}$ is some constant vector. This unambiguously fixes

$$\rho(\theta) = -\frac{\mathrm{i}}{6}\varepsilon_{abc}\theta^a\theta^b\theta^c + C^a\theta^a. \tag{11.14}$$

This all sounds very reasonable, yet the rather obvious requirement

$$\langle f^* f \rangle \geqslant 0 \tag{11.15}$$

is not met, as is clear from the examples

$$f_\pm = \frac{\theta^1 \mp \mathrm{i}\theta^2}{\sqrt{2}},$$

which yield

$$\langle f^*_\pm f_\pm \rangle = \mp \int C_3\theta^1\theta^2\theta^3\mathrm{d}\theta^3\mathrm{d}\theta^2\mathrm{d}\theta^1 = \mp C_3,$$

one of which must be negative for $C_3 \neq 0$. To insure the inequality (11.15) we would have to set $\vec{C} = 0$, thus entirely trivializing the problem. Remarkably this difficulty is cured upon quantization. To quantize this system we could proceed directly from the action via path integrals. We choose the more familiar Dirac approach of replacing Poisson brackets by – in this case – *graded* commutators:

$$[\ ]_P \to \frac{\mathrm{i}}{\hbar}[\ ]. \tag{11.16}$$

Thus in particular, upon quantization equation (11.8) becomes

$$[\theta^a, \theta^b]_+ = \hbar\delta^{ab}. \tag{11.17}$$

It is now convenient to define $\sigma_a = (2/\hbar)^{1/2}\theta^a$ whereupon the σ_as obey the Clifford algebra defining relations

$$[\sigma_a, \sigma_b] = 2\delta_{ab}; \tag{11.18}$$

they are just the Pauli matrices. Quantization sends the Grassmann algebra into a Clifford algebra. The earlier defined spin vector $\vec{S} = -\frac{1}{2}\mathrm{i}\vec{\theta} \times \vec{\theta}$ now becomes

$$\vec{S} = -\tfrac{1}{2}\mathrm{i}\tfrac{1}{2}\hbar\vec{\sigma} \times \vec{\sigma} = \tfrac{1}{2}\hbar\vec{\sigma}, \tag{11.19}$$

and its components obey the standard commutation relations

$$[S_a, S_b] = \mathrm{i}\hbar\varepsilon_{abc}S_c. \tag{11.20}$$

In the standard approach one starts from these commutation relations, and discovers that they have a spinorial representation in terms of Pauli matrices, which also 'happen' to obey the anticommutation relations (11.18). Here just the reverse happens. The Dirac quantization (11.16) yields directly these anti-commutation relations, and the $\mathfrak{su}(2)$ algebra of angular momentum is a by-product.

Finally, in quantum theory the density function $\rho(\theta)$ gets replaced to within a factor $(\frac{1}{2}\hbar)^{3/2}$ by the density matrix

$$\boldsymbol{\rho}(\theta) = 2\left(\tfrac{1}{2} + \vec{C}\frac{\vec{\sigma}}{\hbar}\right) \tag{11.21}$$

whereas

$$\mathrm{i}\int f\rho\,\mathrm{d}\theta \to \mathrm{Tr}(\boldsymbol{\rho} f). \tag{11.22}$$

Now $\boldsymbol{\rho}$ is positive semidefinite (i.e., the inequality (11.15) holds) provided only

$$|\vec{C}| \leqslant \tfrac{1}{2}\hbar, \tag{11.23}$$

which is certainly physically acceptable.

The full meaning of the 'pseudoclassical limit' should now become clear. As emphasized in the textbooks, spin, unlike orbital angular momentum, *is* fixed for a given type of particle, usually at some low integer multiple of $\hbar/2$. So in the classical limit, as $\hbar \to 0$, spin disappears. Spin is a purely quantum-mechanical phenomenon. Yet, we can contemplate an intermediate case in which $\hbar$ is set to zero, but the spin vectors that normally generate the Pauli–Clifford algebra, are kept, as anticommuting quantities that generate a Grassmann algebra (Casalbuoni 1976, 1976*a*). This Grassmann algebra is a contraction of the Clifford algebra as $\hbar \to 0$, in a sense not dissimilar from the Galilei algebra being the contraction of the Poincaré algebra as the inverse of the velocity of light goes to zero. The θs, the Grassmann variables of the pseudoclassical action embody this remanent spin at $\hbar \to 0$. They are not zero as in classical mechanics, but they square to zero.

Pseudoclassical mechanics answers an important question. Take an ordinary classical point particle. Quantize by path integral or by canonical techniques. What you obtain is a first quantized picture governed by Schrödinger's wave equation. But what is the classical system which when quantized yields Pauli's two-component wave equations for a non-relativistic electron? Ordinary classical mechanics has no answer to this question; hence all the statements about spin being a purely quantum phenomenon. The answer is given by pseudoclassical mechanics. The system described by the pseudoclassical action (11.11) when quantized, yields Pauli's nonrelativistic electron.

Evidently the next question is whether we can find the pseudoclassical action which upon quantization yields Dirac's equation for the relativistic electron. We can, as we now show. The idea is to append to the position 4-vector q^μ, four anticommuting quantities θ^μ, which upon quantization will grow into the four Dirac matrices γ^μ, just as the nonrelativistic θs grew into Pauli matrices. We start from the Lorentz and reparametrization invariant action (Berezin & Marinov 1977, Brink, Deser, Zumino, di Vecchia & Howe 1976)

$$S=\int \mathrm{d}\tau\{-m(-\dot{q}^2)^{1/2}+\tfrac{1}{2}\mathrm{i}[\theta\dot{\theta}+(u\theta)(u\dot{\theta})]\} \tag{11.24}$$

where $u=\dot{q}/(-\dot{q}^2)^{1/2}$ and τ is a *Bose* parameter. The bilinear form $\theta\dot{\theta}+(u\theta)(u\dot{\theta})$ is degenerate, so this action does not determine $u\dot{\theta}$. In the usual way we impose a constraint

$$u\theta+\theta_5=0 \tag{11.25}$$

(with θ_5 Fermi), and add to the lagrangian the constraint multiplied by a fermionic lagrange multiplier λ. So now

$$S=\int \mathrm{d}\tau\{-m(-\dot{q}^2)^{1/2}+\tfrac{1}{2}\mathrm{i}[\theta\dot{\theta}+\theta_5\dot{\theta}_5-(u\theta+\theta_5)\lambda]\}. \tag{11.26}$$

The momentum is given by

$$p^\mu=\frac{\partial\mathscr{L}}{\partial\dot{q}_\mu}=mu^\mu-\tfrac{1}{2}\mathrm{i}[\theta^\mu+(u\theta)u^\mu]\frac{\lambda}{(-\dot{q}^2)^{1/2}} \tag{11.27}$$

so that (using $\lambda^2=0$ which follows from the Fermi nature of λ)

$$p^2=m^2u^2-\frac{2\mathrm{i}}{2}m[u\theta+(u\theta)u^2]\frac{\lambda}{(-\dot{q}^2)^{1/2}}=-m^2 \tag{11.28a}$$

since $u^2=-1$. Thus

$$p^2+m^2=0. \tag{11.28b}$$

Similarly,

$$p\theta = mu\theta - \tfrac{1}{2}\mathrm{i}[-\theta\theta - (u\theta)(u\theta)]\frac{\lambda}{(-\dot{q}^2)^{1/2}} = m(u\theta) = -m\theta_5 \quad (11.29)$$

where the Fermi nature of θ^μ has been taken into account. We now have the constraints

$$p\theta + m\theta_5 = 0 \quad (11.30a)$$

$$p^2 + m^2 = 0. \quad (11.30b)$$

Upon quantization, the constraints are to be imposed on all physical states $|\psi\rangle$

$$(p\theta + m\theta_5)|\psi\rangle = 0 \quad (11.31a)$$

$$(p^2 + m^2)|\psi\rangle = 0 \quad (11.31b)$$

and, just as before, the Poisson brackets of the θs and of θ_5 are to be converted into anticommutators ($\eta_{\mu\nu} \equiv$ Minkowski metric):

$$[\theta_\mu, \theta_\nu]_+ = \hbar\eta_{\mu\nu} \quad [\theta_5, \theta_5]_+ = \hbar \quad [\theta_\mu, \theta_5]_+ = 0. \quad (11.32)$$

Defining

$$\gamma_5\gamma_\mu = \left(\frac{2}{\hbar}\right)^{1/2}\theta_\mu \quad \gamma_5 = \left(\frac{2}{\hbar}\right)^{1/2}\theta_5 \quad (11.33)$$

the γs obey their ordinary Dirac–Clifford algebra and the first constraint (11.31) becomes the Dirac equation, as expected. Including an external electromagnetic field in the action, one can derive at the pseudoclassical level the equations for the spin precession (Bargmann, Michel & Telegdi 1959). Similar techniques work for the massless case. For higher spin it is simplest to start from the Dirac-like equations of Bargmann & Wigner (1948), which essentially view the higher spin as built from spin one-half pieces. A spin j wave-function has $2j$ spinorial indices each acted on by its own γ-matrices. One thus introduces at the pseudoclassical level $2j$ Grassmann four vectors. Of course $2j$ is odd (even) for j half-odd-integer (integer), which makes one wonder how much of the spin–statistics connection could already by encoded at the pseudoclassical level, even if one believes that the full connection can only emerge in relativistic quantum field theory.

It should be added that the action (11.24), which upon quantization yielded Dirac's equation is by no means unique. There exists an alternative formulation in which the Grassmann degrees of freedom are chosen spinorial at the pseudoclassical level (Freund in Ferber 1978, Brink & Schwarz 1981).

12

Supersymmetric field theories in two space–time dimensions

While this time the end product will be a genuine field theory with infinitely many degrees of freedom, the construction follows quite closely the pattern described in chapter 10. We consider Majorana spinors θ^α ($\alpha = 1, 2$) which have two real components. Correspondingly the $N = 1$ superfields will admit three types of terms in their expansion (like the $N = 2$ case in one time, zero space dimensions). In repeating the steps leading to equation (10.19), we will occasionally need to reorder the spinors in a product. This is accomplished via Fierz rearrangements. Generically, the following Fierz rearrangement will be all we need:

$$\bar{\theta}_\alpha \varphi_\alpha \theta_\beta = -\tfrac{1}{2}\bar{\theta}_\alpha \theta_\alpha \varphi_\beta. \tag{12.1}$$

Its proof is outlined in the following sequence of equalities

$$\bar{\theta}_\alpha \varphi_\alpha \begin{pmatrix} \theta_1 \\ \theta_2 \end{pmatrix} = (\theta_1 \ \ \theta_2) \begin{pmatrix} 0 & 1 \\ -1 & 0 \end{pmatrix} \begin{pmatrix} \varphi_1 \\ \varphi_2 \end{pmatrix} \begin{pmatrix} \theta_1 \\ \theta_2 \end{pmatrix} = (\theta_1 \varphi_2 - \theta_2 \varphi_1) \begin{pmatrix} \theta_1 \\ \theta_2 \end{pmatrix}$$

$$= -\begin{pmatrix} \theta_1 \theta_2 \varphi_1 \\ \theta_1 \theta_2 \varphi_2 \end{pmatrix} = -\tfrac{1}{2}\bar{\theta}\theta \begin{pmatrix} \varphi_1 \\ \varphi_2 \end{pmatrix}$$

where we used the anticommutativity of the Fermi θ_is and φ_is and the Majorana spinor relations

$$\bar{\theta} = \theta^{\mathrm{T}} \gamma^0, \quad \theta = \theta^*, \quad \theta = -(\bar{\theta}\gamma^0)^{\mathrm{T}}, \quad \gamma^0 = \begin{pmatrix} 0 & 1 \\ -1 & 0 \end{pmatrix}, \quad \tfrac{1}{2}\bar{\theta}\theta = \theta_1 \theta_2. \tag{12.2}$$

The superfield $\phi(x, \theta)$ is then

$$\phi(x, \theta) = A(x) + \mathrm{i}\bar{\theta}\psi(x) + \tfrac{1}{2}\mathrm{i}\bar{\theta}\theta F(x) \tag{12.3}$$

since all terms of higher order in θ, in the expansion of type (8.2), vanish. The supersymmetry generators are given by

$$Q_\alpha = \frac{\partial}{\partial \bar{\theta}^\alpha} - \mathrm{i}(\gamma^\mu \theta)_\alpha \partial_\mu \tag{12.4}$$

and the transformation law of ϕ is

$$\delta\phi = \bar{\alpha}Q\phi = -\mathrm{i}\bar{\alpha}\gamma^{\mu}\theta\partial_{\mu}A + \mathrm{i}\bar{\alpha}\psi + \bar{\alpha}\gamma^{\mu}\theta\bar{\theta}\partial_{\mu}\psi + \mathrm{i}\bar{\alpha}\theta F, \tag{12.5a}$$

just as in (10.11*a*). From here the transformation laws of the component fields

$$\delta\phi = \delta A + \mathrm{i}\bar{\theta}\delta\psi + \tfrac{1}{2}\mathrm{i}\bar{\theta}\theta\delta F$$

are obtained as

$$\left.\begin{aligned} \delta A &= \mathrm{i}\bar{\alpha}\psi \\ \delta\psi &= (\gamma\partial A + F)\alpha \\ \delta F &= \mathrm{i}\bar{\alpha}\gamma\partial\psi. \end{aligned}\right\} \tag{12.5b}$$

In going from (12.5*a*) to (12.5*b*) we performed Fierz rearrangements in the first and third terms of equation (12.5*a*), we used the Fermi nature of $\bar{\alpha}$ and the second and third of the following set of two-dimensional Dirac–Majorana algebra relations

$$\begin{gathered} C = \gamma^0 = \mathrm{i}\sigma_2 \quad C^{\mathrm{T}} = -C = C^{-1} \quad C\gamma^{\nu\mathrm{T}}C^{-1} = -\gamma^{\nu} \\ \mathrm{Tr}\,C^2 = -2 \quad \mathrm{Tr}\,C\gamma^{\nu}C = 0. \end{gathered} \tag{12.6}$$

The covariant derivative is the spinor

$$D_{\alpha} = \frac{\partial}{\partial\bar{\theta}^{\alpha}} + \mathrm{i}(\gamma^{\mu}\theta)_{\alpha}\partial_{\mu} \tag{12.7}$$

and as before has the properties

$$[D_{\alpha}, Q_{\beta}] = 0 \quad [\bar{D}D, Q_{\alpha}] = 0 \tag{12.8}$$

The invariance under supersymmetry of the lagrangian

$$\mathscr{L}_0 = \tfrac{1}{4}\phi^{+}\bar{D}D\phi \tag{12.9}$$

is proved exactly as in chapter 10. To expand $\mathscr{L}_0$ in components we need the two-dimensional Dirac algebra relations (12.6). We start by calculating the field equation superfield (using equations (12.3), (12.6), (12.2), (12.7))

$$\begin{aligned} \bar{D}D\phi &= D_{\alpha}C_{\alpha\beta}D_{\beta}\phi = -C_{\beta\alpha}D_{\alpha}D_{\beta}\phi \\ &= -C_{\beta\alpha}\left[\frac{\partial}{\partial\bar{\theta}^{\alpha}} + \mathrm{i}(\bar{\theta}\gamma^{\mu}C)_{\alpha}\partial_{\mu}\right]\left[\frac{\partial}{\partial\bar{\theta}^{\beta}} + \mathrm{i}(\bar{\theta}\gamma^{\nu}C)_{\beta}\partial_{\nu}\right](A + \mathrm{i}\bar{\theta}\psi + \tfrac{1}{2}\mathrm{i}\bar{\theta}\theta F) \\ &= -C_{\beta\alpha}[-\mathrm{i}C_{\alpha\beta}F + (\bar{\theta}\gamma^{\nu}C)_{\beta}\partial_{\nu}\psi_{\alpha} - (\bar{\theta}\gamma^{\mu}C)_{\alpha}\partial_{\mu}\psi_{\beta} - (\bar{\theta}\gamma^{\mu}C)_{\alpha}\partial_{\mu}F\theta_{\beta} \\ &\quad - (\bar{\theta}\gamma^{\mu}C)_{\alpha}(\bar{\theta}\gamma^{\nu}C)_{\beta}\partial_{\mu}\partial_{\nu}A] \\ &= 2(-\mathrm{i}F + \bar{\theta}\gamma\partial\psi + \tfrac{1}{2}\bar{\theta}\theta\,\Box A) \end{aligned} \tag{12.10}$$

where

$$\Box A = \eta^{\mu\nu}\partial_\mu\partial_\nu A = (-\partial_0^2 + \partial_1^2)A \tag{12.11}$$

and the term containing $\partial_\mu F$ drops out on account of $\bar{\theta}\gamma_\mu\theta = 0$. To get to the action we still multiply by ϕ and collect all the terms proportional to $\bar{\theta}\theta$, the only ones that survive the θ^1 and θ^2 integrations. In this process we encounter the term $(\bar{\theta}\mathrm{i}\gamma\partial\psi)(\bar{\psi}\theta)$ which has to be Fierz transformed according to the rule (12.1) into $-\frac{1}{2}(\bar{\theta}\theta)(\bar{\psi}\mathrm{i}\gamma\partial\psi)$. The rest is straightforward and using $\frac{1}{2}\bar{\theta}\theta = \theta_1\theta_2$ we find

$$S_0 = \int \mathrm{d}^2x\mathrm{d}^2\theta\mathscr{L}_0 = -\tfrac{1}{2}\int \theta_1\theta_2\mathrm{d}\theta_2\mathrm{d}\theta_1 \int \mathrm{d}^2x(-A\Box A + \bar{\psi}\mathrm{i}\gamma\partial\psi - F^2) \tag{12.12}$$

so that, up to a surface term,

$$S_0 = -\tfrac{1}{2}\int \mathrm{d}^2x(\partial A\partial A + \mathrm{i}\bar{\psi}\gamma\partial\psi - F^2). \tag{12.13}$$

As a check, this component field action is indeed invariant up to a surface term under the component supersymmetry transformations (12.5*b*). Again F is a nonpropagating auxiliary field, whereas A and ψ are respectively a free massless Bose scalar and a free massless Fermi–Majorana spinor, each with one physical degree of freedom.

At this point we may inquire into the role of the nonpropagating auxiliary field F. Its 'field equation' is $F = 0$, so F really 'cuts no ice'. So, why not set $F = 0$ in the action (12.13) and be done with it? The point is that F actively participates in the supersymmetry transformation laws (12.5). Were one to set $F = 0$ in the action (12.13), the so truncated action would continue to be invariant up to a surface term under the truncated supersymmetry transformations

$$\left.\begin{aligned} \delta A &= \mathrm{i}\bar{\alpha}\psi \\ \delta\psi &= \gamma\partial A\alpha \end{aligned}\right\} \tag{12.14}$$

obtained from (12.5*b*) by setting $F = 0$. Yet these truncated infinitesimal supersymmetry transformations (12.14) do not close unless we impose the field equations (or mass-shell conditions) $\mathrm{i}\gamma\partial\psi = 0$, $\Box A = 0$, on the propagating fields ψ and A. In other words, the commutator of two infinitesimal transformations of the type (12.14) yields a translation only on-shell!

Another way of looking at this matter is to note that in two space–time dimensions the Majorana fermion ψ has one degree of freedom on-shell, i.e., after imposition of the Dirac equation. So does the propagating scalar field A. In a supersymmetric theory, as we saw in chapter 5, the numbers of Bose and Fermi degrees of freedom must be the same, so on-shell A and ψ are all one needs. But as is well known, the Dirac equation cuts by a factor two the number of spinor degrees of freedom. Going off-shell, i.e., to ψ-fields that do not obey the Dirac equation, thus doubles the number of Fermi degrees of freedom, while the scalar field A, even off-shell has still only one degree of freedom. To match the Bose and Fermi degrees of freedom, *off-shell* as well, we must call upon another Bose degree of freedom. The auxiliary scalar field F has the role of providing this needed degree of freedom off-shell, while automatically vanishing on-shell just as the extra off-shell spinorial degree of freedom. We have given this discussion in full detail in this simple case. Throughout supersymmetry theory we will encounter such auxiliary fields, each time making a discrete disappearing act on shell, but asserting their presence off-shell where they are needed to restore the Fermi–Bose balance required by supersymmetry.

As in one dimension we can now add an interaction in the form

$$S_\mathrm{I} = \mathrm{i}\int \mathrm{d}^2x\,\mathrm{d}^2\theta\, W(\phi), \tag{12.15}$$

where the simplest nontrivial superspace potential

$$W(\phi) = -(\tfrac{1}{3}g\phi^3 + \lambda\phi) \tag{12.16}$$

yields

$$S_\mathrm{I} = \int \mathrm{d}^2x[\lambda F + g(FA^2 - \mathrm{i}\bar{\psi}\psi A)]. \tag{12.17}$$

The field equation for F is still algebraic:

$$F = -(\lambda + gA^2), \tag{12.18}$$

which when inserted in the action yields

$$S = S_0 + S_\mathrm{I} = \int \mathrm{d}^2x[-\tfrac{1}{2}\partial A\partial A - \tfrac{1}{2}\mathrm{i}\bar{\psi}\gamma\partial\psi - \tfrac{1}{2}(\lambda + gA^2)^2 - \mathrm{i}g\bar{\psi}\psi A]. \tag{12.19}$$

Now the scalar field exhibits a quartic self-interaction and a Yukawa interaction with the fermion. The effective potential at the tree level is

$$\tilde{V} = \tfrac{1}{2}F^2 = \tfrac{1}{2}(\lambda + gA^2)^2. \tag{12.20}$$

For λ/g negative, $\tilde{V}$ has the familiar double well shape with two minima

$\tilde{V}_{\min}=0$ at $\langle A\rangle = \pm A_0 \equiv \pm(-\lambda/g)^{1/2}$, corresponding to $\langle F\rangle = 0$ and $m_A = m_\psi = -2(-g\lambda)^{1/2}$. The fields A and ψ have equal masses, supersymmetry is exact at the tree level but the reflection symmetry $A \to -A$ is broken. In the next chapter we shall see whether this can be maintained in quantum theory. For λ/g positive, $\tilde{V}$ has a single minimum $\tilde{V}_{\min} = \lambda^2/2$ at $\langle A\rangle = 0$. The $A \to -A$ reflection symmetry is observed now, but supersymmetry is broken at the tree level: $m_A = (2g\lambda)^{1/2}$, $m_\psi = 0$, $\langle F\rangle = \lambda$. The vanishing ψ-mass signals a *Nambu–Goldstone fermion* sometimes also called a *goldstino*, associated with this spontaneous supersymmetry breaking at the tree level. This is also reflected in the transformation law $\delta\psi = \lambda\alpha =$ constant, which is obtained from (12.5*b*) for $F = \langle F\rangle = \lambda$ and $A = 0$.

In the next chapter we address ourselves to this phenomenon of spontaneous supersymmetry breaking.

13

Spontaneous supersymmetry breaking

To get an idea about supersymmetry breaking, it is instructive to start with the theories in one time and zero space dimensions discussed in chapter 10 (Witten 1982). These have a finite number of degrees of freedom and yet can exhibit spontaneous supersymmetry breaking. For such theories we had the supersymmetry algebra (see equations (10.3))

$$Q^2 = -H, \quad [H,H] = [H,Q] = 0 \tag{13.1}$$

Consequently, for any state, $|s\rangle$, given that Q is hermitean, we have

$$\langle s|H|s\rangle = \| Q|s\rangle \|^2 \geqslant 0 \tag{13.2}$$

and energy is positive semidefinite. Thus, the lowest energy a state can have, is zero. Any zero energy state is then guaranteed to be a possible ground state.

On the other hand, a state is supersymmetric if it is annihilated by the generator Q of supersymmetry. But equation (13.2) then implies that any state of zero energy is supersymmetric and vice versa. Thus any supersymmetric quantum system, must have a zero energy state (which then functions as a supersymmetric ground state).

To decide whether a system is supersymmetric, or breaks supersymmetry spontaneously, we just have to find out whether it has zero energy states. To this end it is convenient to introduce Witten's operator $(-1)^F$ with the properties

$$(-1)^F|f\rangle = -|f\rangle \quad (-1)^F|b\rangle = |b\rangle \tag{13.3}$$

for any Fermi state $|f\rangle$ and Bose state $|b\rangle$ in the system's Hilbert space. By the spin–statistics connection, in higher dimensions $(-1)^F$ can be realized in terms of the generator of rotations J_z around say the z-axis as

$$(-1)^F = \exp(\mathrm{i}2\pi J_z) \tag{13.4}$$

The trace of $(-1)^F$ contains information about zero energy states. To see this, let $|b\rangle$ be any bosonic energy eigenstate of energy $E > 0$. The action of Q on $|b\rangle$ produces a fermionic state which we can write as

$$Q|b\rangle = E^{1/2}|f\rangle \quad (E^{1/2} \neq 0). \tag{13.5a}$$

Then

$$Q|f\rangle = \frac{Q^2}{E^{1/2}}|b\rangle = \frac{H}{E^{1/2}}|b\rangle = E^{1/2}|b\rangle, \tag{13.5b}$$

so that the two states $|b\rangle$ and $|f\rangle$ are paired in an irreducible multiplet of the supersymmetry algebra. A zero energy bosonic state $|b_0\rangle$ is supersymmetric as we have seen,

$$Q|b_0\rangle = 0. \tag{13.6a}$$

Similarly for a zero energy fermionic state $|f_0\rangle$

$$Q|f_0\rangle = 0. \tag{13.6b}$$

Thus the state $|b_0\rangle$ is by itself a supersymmetry singlet, as is $|f_0\rangle$. We see that for finite energy the number of bosonic states equals that of the fermionic states, but for zero energy states the numbers of bosonic and fermionic states need not be equal. Denote by $\nu(E)$ the difference between the number of bosonic and fermionic states of energy E in the system. Then the observation just made means that $\nu(E)$ must vanish at all energies except at zero energy where it equals some integer.

Were we now to vary the parameters of the theory (masses, coupling constants,...) the energies of the states would vary accordingly. The validity of the supersymmetry algebra requires though, that the states move around in Bose–Fermi pairs, otherwise there would appear nonvanishing energy values with an excess of Bose or Fermi states. In particular, states land on, or leave zero energy in such Bose–Fermi pairs, so that not only does $\nu(E)$ vanish for $E \neq 0$ at all values of the parameters, but even the integer value of $\nu(0)$ is frozen in, as we vary the parameters. But if $\nu(0) \neq 0$ then there must exist zero energy states in the system and therefore supersymmetry does not break spontaneously. If $\nu(0) = 0$ all we know is that even at $E = 0$ the number of bosonic and fermionic states are equal, but that by itself is not very instructive.

Since every bosonic (fermionic) state contributes $+1$ (-1) to the trace of Witten's operator $\mathrm{tr}(-1)^F$, it follows (on account of the Bose–Fermi pairing of $E \neq 0$ states) that

$$\mathrm{tr}(-1)^F = \nu(0). \tag{13.7}$$

The trace being taken over a Hilbert space, questions of convergence arise, but can be readily dealt with by writing $\lim_{\beta \to 0^+} \mathrm{tr}[(-1)^F \exp(-\beta H)]$ on the left-hand side of equation (13.7). The important thing is that the value of $\mathrm{tr}(-1)^F$ is impervious to variations of the parameters as long as these do not go to zero, or no new parameters are introduced. To illustrate these ideas we discuss two simple examples.

Example 1: The $N = 2$ superpoint particle in $d = 1$ space dimensions.
In chapter 10 we had the action (see equations (10.24))

$$S = \int \mathrm{d}t \left[\tfrac{1}{2}\dot{x}^2 + \tfrac{1}{2}\mathrm{i}\theta_\alpha\dot{\theta}_\alpha - \frac{1}{2}\left(\frac{\mathrm{d}W}{\mathrm{d}x}\right)^2 + \mathrm{i}\frac{\mathrm{d}^2W}{\mathrm{d}x^2}\theta_1\theta_2 \right] \tag{13.8}$$

(the Latin indices of equations (10.24) have been suppressed since for $d = 1$ they only take the value 1). The corresponding hamiltonian (see also chapter 11) is

$$H = \tfrac{1}{2}p^2 + \frac{1}{2}\left(\frac{\mathrm{d}W}{\mathrm{d}x}\right)^2 - \mathrm{i}\frac{\mathrm{d}^2W}{\mathrm{d}x^2}\theta_1\theta_2. \tag{13.9}$$

For simplicity we choose

$$W(x) = \tfrac{1}{3}x^3 + ax \tag{13.10}$$

so that

$$H = \tfrac{1}{2}p^2 + \tfrac{1}{2}(x^2 + a)^2 - 2\mathrm{i}x\theta_1\theta_2 \tag{13.11}$$

Here

$$[\theta_\alpha, \theta_\beta]_+ = 0 \quad \text{for } \alpha, \beta = 1, 2. \tag{13.12}$$

Upon quantization this is replaced (see equation (11.17)) by

$$[\theta_\alpha, \theta_\beta]_+ = \hbar\delta_{\alpha\beta}, \tag{13.13}$$

θ_α can be represented by Pauli matrices

$$\theta_\alpha = (\tfrac{1}{2}\hbar)^{1/2}\sigma_\alpha \quad \alpha = 1, 2 \tag{13.14}$$

and

$$H = \tfrac{1}{2}p^2 + \tfrac{1}{2}(x^2 + a)^2 + \hbar\sigma_3 x \tag{13.15}$$

At the classical (or tree) level: $\hbar \to 0$, so that the ground state energy is determined by the minimum of the potential $(x^2 + a)^2$. For $a > 0$, $V(x) \geqslant a^2/2 > 0$ and supersymmetry is spontaneously broken. For $a < 0$, V attains the minimum value $V_{\text{min}} = 0$ at $x = \pm(|a|)^{1/2}$ and it would appear that supersymmetry is unbroken. This is however a tree level result that does not hold in the full quantum theory. To see this, consider the transformation

$$H \to \tilde{H} = \mathrm{e}^{2ax}H\mathrm{e}^{-2ax} \tag{13.16}$$

This transformation is so chosen that $\tilde{H}$ is precisely H with the sign of the parameter a inverted. Under this transformation the two supersymmetry generators

$$Q_1 = \frac{1}{\sqrt{2}}[\sigma_1 p + \sigma_2(x^2 + a)]$$
$$Q_2 = \frac{1}{\sqrt{2}}[\sigma_2 p - \sigma_1(x^2 + a)] \tag{13.17}$$

transform as

$$Q_\alpha \to \tilde{Q}_\alpha = e^{2ax} Q_\alpha e^{-2ax} \tag{13.18}$$

It is readily checked that the expressions given here for Q_α obey the $N=2$ supersymmetry algebra $[Q_\alpha, Q_\beta] = 2\delta_{\alpha\beta}H$. The operator e^{2ax} not being unitary, the theory based on $\tilde{Q}_\alpha$ and $\tilde{H}$ is not equivalent to that based on Q_α and H. Still, the number of zero energy states of H and of $\tilde{H}$ are the same, as we shall prove below. Yet for the theory based on H we have already established the absence of zero energy states, which in turn implies spontaneous supersymmetry breakdown. Thus in the theory with $a<0$, described by $\tilde{H}$, tree level appearances notwithstanding, supersymmetry is also spontaneously broken.

To complete the argument we have to prove the equality of the numbers of zero energy states in the H and $\tilde{H}$ theories. Consider the operators $Q_\pm = (1/\sqrt{2})(Q_1 \pm iQ_2)$. The supersymmetry algebra imposes

$$Q_\pm^2 = 0 \quad [Q_+, Q_-] = H. \tag{13.19}$$

A state $|s\rangle$ for which $Q_+|s\rangle = 0$ will be called a *closed state.* A state $|s\rangle$, for which there exists a state $|t\rangle$ such that $|s\rangle = Q_+|t\rangle$, will be called an *exact state* (this nomenclature is of cohomological origin). By virtue of $Q_+^2 = 0$, every exact state is closed. We now ask whether there exist closed states which are not exact. We start from a basis $|s_E\rangle$ of hamiltonian eigenstates

$$H|s_E\rangle = E|s_E\rangle. \tag{13.20}$$

We first show that every *closed* state $|s_E\rangle$ corresponding to a nonvanishing eigenvalue $E \neq 0$ is exact. To this end, consider the state $|t_E\rangle = (1/E)Q_-|s_E\rangle$. We have $Q_+|t_E\rangle = (1/E)Q_+Q_-|s_E\rangle = (1/E)\ [Q_+, Q_-]|s_E\rangle$ (since $Q_-Q_+|s_E\rangle = 0$, $|s_E\rangle$ being closed by assumption). Using the supersymmetry algebra (13.19), we further find $Q_+|t_E\rangle = (1/E)H|s_E\rangle$, thus proving $|s_E\rangle$ to be exact. If however $E=0$, the supersymmetry algebra relation $[Q_+, H] = 0$ implies that, if there exists a state $|t_0\rangle$ for which $|s_0\rangle = Q_+|t_0\rangle$, then $H|t_0\rangle = 0$. But Q_α, and therefore Q_+ annihilate all zero energy states (i.e., all zero energy states are closed), so that $Q_+|t_0\rangle = 0$ and therefore $|s_0\rangle = Q_+|t_0\rangle$ cannot hold. Thus the *zero energy states are all the closed states which are not exact.* Under a similarity transformation M, Q_+ changes

$$Q_+ \rightarrow \tilde{Q}_+ = M^{-1}Q_+M. \tag{13.21}$$

Consider a zero energy state $|s_0\rangle$ of H. Under the action of Q_+ it is closed but not exact. The state $|\tilde{s}_0\rangle = M^{-1}|s_0\rangle$ is then closed but not exact under the action of $\tilde{Q}_+$ ($\tilde{Q}_+|\tilde{s}_0\rangle = M^{-1}Q_+MM^{-1}|s_0\rangle = M^{-1}Q_+|s_0\rangle = 0$; if $|\tilde{s}_0\rangle$ were exact, then from the existence of $|\tilde{t}_0\rangle$ such that $|\tilde{s}_0\rangle = \tilde{Q}_+|\tilde{t}_0\rangle$, we could conclude $|s_0\rangle = M|\tilde{s}_0\rangle = MM^{-1}Q_+M|\tilde{t}_0\rangle = Q_+(M|\tilde{t}_0\rangle)$, which would mean $|s_0\rangle$ is exact contrary to the assumption). Thus under a similarity transformation zero energy states of H map into zero energy states of $\tilde{H}$. The converse is then also easily established, so that the similarity transformation provides a one-to-one map between the zero energy states of H and those of $\tilde{H}$. Hence the numbers of zero energy states of H and $\tilde{H}$ are equal as claimed. For the case of the supersymmetric point particle all that needs to be added is that the transformation (13.16) is a similarity.

Example 2: Supersymmetric two-dimensional field theory
Consider the theory discussed in chapter 12, with action (12.19). There we have discussed supersymmetry breaking at the tree level for this theory. But as we saw in the previous example such discussions may be misleading. Indeed, were we to treat this model in a finite volume, the sign of λ/g could then be changed by a similarity transformation, and just as in example 1 we would have spontaneous supersymmetry breaking for either sign of λ/g. For negative λ/g, as was already noted in chapter 12, $\langle A\rangle = \pm A_0 \neq 0$ so that $m_\psi \neq 0$. There is no Nambu–Goldstone fermion, and there can be no spontaneous supersymmetry breaking in the infinite volume limit. In this limit Poincaré supersymmetry is established, and in a Poincaré superinvariant theory, if a symmetry or supersymmetry spontaneously breaks, then a Nambu–Goldstone particle (boson or fermion as the case may be) *must* appear. This is the contents of Goldstone's theorem (Goldstone, Salam & Weinberg 1962), which ceases to hold in a finite volume or for nonrelativistic point particles. Thus, in the infinite volume limit, the discussion of chapter 12 remains valid. In this theory $\mathrm{tr}(-1)^F$ vanishes for both signs of λ/g. For $\lambda/g > 0$ there are no zero energy states, for $\lambda/g < 0$ there are two such states.

Much of our discussion was based on the observation, made at the beginning of this chapter that in a supersymmetric ground state the energy must vanish on account of the supersymmetry algebra (13.1). This observation has a close relativistic counterpart (Zumino 1975) which we now present.

In a relativistic quantum field theory consider the vacuum expectation $\langle 0|T_{\mu\nu}(x)|0\rangle$ of the energy momentum tensor $T_{\mu\nu}(x)$. Generically $\langle 0|T_{\mu\nu}(x)|0\rangle$ is set equal to zero through subtractions performed order-by-order in perturbation theory. In a supersymmetric quantum field theory the Poincaré superalgebra implies

$$[Q_\alpha, Q_\beta] = 2(\gamma^\mu C)_{\alpha\beta}P_\mu \tag{13.22}$$

where Q_α is the space integral (at constant time) of the time-like component of a fermionic 'supercurrent' S^α_μ and P_μ the similar integral of the component $T_{\mu 0}$ of the energy–momentum tensor. Thus dropping the space integration we find

$$[Q_\alpha, S_{0\mu}(x)] = 2(\gamma^\mu C)_{\alpha\beta}T_{\mu 0}(x) + \mathrm{ST}, \tag{13.23}$$

where ST stands for a Schwinger term. In a supersymmetric theory $Q_\alpha|0\rangle = 0$ and $\langle 0|Q_\alpha = 0$, so that taking the vacuum expectation value of equation (13.23) and recalling $\langle 0|\mathrm{ST}|0\rangle = 0$, we see that $\langle 0|T_{\mu 0}|0\rangle = 0$, or because of Lorentz invariance

$$\langle 0|T_{\mu\nu}|0\rangle = 0 \tag{13.24}$$

For free fields, viewed as a collection of harmonic oscillators, what happens is, that the positive zero point energy of the Bose oscillators cancels against the negative zero point energy of the equally numerous Fermi oscillators. Of course, in view of our discussion above, (13.24) breaks down if supersymmetry is spontaneously broken.

14

Vector and chiral superfields in four-dimensional space–time

In the case of four-dimensional space–time the simplest, $N = 1$, superspace is $B_L^{4,4}$ involving one Majorana spinor θ^α $(\alpha = 1, 2, 3, 4)$ along with the Minkowski coordinates x^μ $(\mu = 0, 1, 2, 3)$. A superfield $\phi(x^\mu, \theta^\alpha)$ now admits a θ-expansion of the type

$$\phi(x, \theta) = \phi(x) + \phi_\alpha \theta^\alpha + \cdots + \phi_{[\alpha\beta\gamma\delta]} \theta^\alpha \theta^\beta \theta^\gamma \theta^\delta.$$

It is both convenient and conventional to switch to the Weyl notation as explained in chapter 5. We trade-in the real four-component Majorana spinor θ^α for a two-component complex Weyl spinor also called θ^α $(\alpha = 1, 2)$ and its conjugate $\bar{\theta}_{\dot{\alpha}} = (\theta_\alpha)^*$. In what has, by now, become a conventional notation, the θ-expansion is written as

$$\begin{aligned}\phi(x, \theta, \bar{\theta}) = {} & C + \mathrm{i}\theta^\alpha \chi_\alpha - \mathrm{i}\bar{\theta}_{\dot{\alpha}} \bar{\chi}^{\dot{\alpha}} + \theta^\alpha \theta_\alpha \tfrac{1}{2}\mathrm{i}(M + \mathrm{i}N) - \bar{\theta}_{\dot{\alpha}} \bar{\theta}^{\dot{\alpha}} \tfrac{1}{2}\mathrm{i}(M - \mathrm{i}N) \\ & - \theta^\alpha (\sigma_\mu)_{\alpha\beta} \bar{\theta}^\beta V^\mu + \mathrm{i}\theta^\alpha \theta_\alpha \bar{\theta}_\beta (\bar{\lambda}^\beta - \tfrac{1}{2}\mathrm{i}\sigma^\mu{}_\gamma{}^\beta \chi^\gamma) \\ & - \mathrm{i}\bar{\theta}_{\dot{\alpha}} \bar{\theta}^{\dot{\alpha}} \theta^\beta (\lambda_\beta + \tfrac{1}{2}\mathrm{i}\sigma^\mu{}_{\beta\dot{\gamma}} \partial_\mu \bar{\chi}^{\dot{\gamma}}) + \theta^\alpha \theta_\alpha \bar{\theta}_\beta \bar{\theta}^\beta \tfrac{1}{2}(D + \Box C),\end{aligned} \tag{14.1}$$

For completeness we recall the Weyl notation $\theta^\alpha = \varepsilon^{\alpha\beta}\theta_\beta$, $\bar{\theta}^{\dot{\alpha}} = \varepsilon^{\dot{\alpha}\dot{\beta}}\bar{\theta}_{\dot{\beta}}$, $\varepsilon^{12} = \varepsilon^{12} = 1$, $\varepsilon^{\alpha\beta} = -\varepsilon^{\beta\alpha}$, $\varepsilon^{\dot{\alpha}\dot{\beta}} = -\varepsilon^{\dot{\beta}\dot{\alpha}}$, $\sigma^0{}_{\dot{\alpha}\beta}$ is the unit matrix, $\sigma^m{}_{\alpha\beta}$, $m = 1$, 2, 3, the three Pauli matrices. Notice that $\frac{1}{2}\mathrm{i}\sigma^\mu{}_{\alpha\beta}\partial_\mu\bar{\chi}^\beta$ transforms like a spinor with an undotted index which explains its appearance in the company of λ. We shall also use the abbreviations $\phi^\alpha\psi_\alpha = \phi\psi$, $\bar{\phi}_{\dot{\alpha}}\bar{\psi}^{\dot{\alpha}} = \bar{\phi}\bar{\psi}$ for any two spinors ϕ and ψ, or $\bar{\phi}$ and $\bar{\psi}$.

In (14.1) the superfield ϕ has been assumed real. Thus the coefficients of θ^α and of $\bar{\theta}^{\dot{\alpha}}$ are each other's complex conjugates, as are the coefficients of $\theta^\alpha\theta_\alpha$ and $\bar{\theta}^{\dot{\alpha}}\bar{\theta}_{\dot{\alpha}}$, etc.... Strictly speaking, the (dimensionally correct: $\dim \theta_\alpha = \frac{1}{2}$, $\dim \partial_\mu = -1$) terms $\frac{1}{2}\mathrm{i}\sigma\partial\bar{\chi}$ and $\frac{1}{2}\Box C$ in the θ-trilinear and θ-quadrilinear terms are not necessary, they have been included to simplify the transformation laws of the component fields. The highest spin in (14.1) corresponds to the vector field V_μ, hence the name *vector superfield* for ϕ as given by this θ-expansion. The supersymmetry generators are now

$$\left.\begin{aligned} Q_\alpha &= \frac{\partial}{\partial\theta^\alpha} - \mathrm{i}\sigma^\mu{}_{\alpha\dot\alpha}\bar\theta^{\dot\alpha}\partial_\mu \\ \bar Q^{\dot\alpha} &= \frac{\partial}{\partial\bar\theta_{\dot\alpha}} - \mathrm{i}\theta^\alpha\sigma^\mu{}_{\alpha\beta}\varepsilon^{\beta\dot\alpha}\partial_\mu \end{aligned}\right\} \tag{14.2}$$

whereas the covariant derivatives are

$$\left.\begin{aligned} D_\alpha &= \frac{\partial}{\partial\theta^\alpha} + \mathrm{i}\sigma^\mu{}_{\alpha\dot\alpha}\bar\theta^{\dot\alpha}\partial_\mu \\ \bar D_{\dot\alpha} &= -\frac{\partial}{\partial\bar\theta_{\dot\alpha}} - \mathrm{i}\theta^\alpha\sigma^\mu{}_{\alpha\dot\alpha}\partial_\mu \end{aligned}\right\} \tag{14.3}$$

The bracketing rules are

$$\left.\begin{aligned} &[Q_\alpha, \bar Q_{\dot\alpha}] = -[D_\alpha, \bar D_{\dot\alpha}] = 2\mathrm{i}\sigma^\mu{}_{\alpha\dot\alpha}\partial_\mu \\ &[Q_\alpha, Q_\beta] = [\bar Q_{\dot\alpha}, \bar Q_{\dot\beta}] = [D_\alpha, D_\beta] = [\bar D_{\dot\alpha}, \bar D_{\dot\beta}] = 0 \end{aligned}\right\} \tag{14.4}$$

whereas all Ds and $\bar D$s anticommute with all Qs and $\bar Q$s as befits covariant derivatives. All this is but a rehash of what we already had in lower dimensions. In four-dimensions, we have one novel feature: we can impose the supersymmetric constraints (Salam & Strathdee 1975)

$$D_\alpha\phi = 0 \tag{14.5a}$$

or

$$\bar D_{\dot\alpha}\phi = 0, \tag{14.5b}$$

thus obtaining superfields with a smaller number of component fields. Superfields obeying the constraints (14.5*a*) or (14.5*b*) are called respectively *right-handed* and *left-handed*, or simply *chiral*. One may wonder about the possibility of such chiral constraints in the one- and two-dimensional cases treated in chapters 10 and 12. In one dimension as in any odd dimension there are no chiral spinors (see chapter 3) and thus no chiral constraint can be imposed. In two-dimensions there exist ⋈ spinors (table 3.4) and a further chiral constraint on the superfield is enforceable and of interest in superstring theory.

To solve, say, the left-handed constraints (14.5*b*) note that

$$\bar D_{\dot\alpha}\theta^\beta = 0 \quad \bar D_{\dot\alpha}(x^k + \mathrm{i}\theta\sigma^\mu\bar\theta) = 0, \tag{14.6}$$

let $\phi_1(x, \theta)$ be a superfield which, while depending on the even elements x^μ of the Grassmann algebra and on the odd elements θ, does *not* depend on the odd elements $\bar{\theta}$. The identities (14.6) then guarantee that the most general solution of the constant (14.5*b*) is

$$\phi(x^\mu, \theta_\alpha, \bar{\theta}_{\dot{\alpha}}) = \phi_1(y^\mu, \theta_\alpha)|_{y^\mu = x^\mu + i\theta\sigma^\mu\bar{\theta}}. \tag{14.7}$$

The superfield ϕ_1, independent of $\bar{\theta}$, has the much more economical θ-expansion

$$\phi_1(x, \theta) = \tfrac{1}{2}(A - iB) + \theta^\alpha\psi_\alpha + \theta^\alpha\theta_\alpha\tfrac{1}{2}(F + iG). \tag{14.8}$$

Note that this chiral superfield is complex. Assuming the superfield ϕ_1 to be superanalytic, we can expand $\phi_1(x^\mu + i\theta\sigma^\mu\bar{\theta}, \theta^\alpha)$ around the point (x^μ, θ^α). This will bring into play a finite number of terms ($i\theta\sigma^\mu\bar{\theta}$ is nilpotent) containing the derivatives (typical of a Taylor expansion) of the fields A, B and ψ appearing in (14.8). This way, from (14.7) and (14.8) we finally obtain the general form of the chirally constrained superfield $\phi(x, \theta, \bar{\theta})$ as

$$\begin{aligned}\phi(x, \theta, \bar{\theta}) = {} & \tfrac{1}{2}(A - iB) + \theta^\alpha\psi_\alpha + \theta^\alpha\theta_\alpha\tfrac{1}{2}(F + iG) + \tfrac{1}{2}i\theta^\alpha\sigma^\mu{}_{\alpha\beta}\bar{\theta}^\beta\partial_\mu(A - iB) \\ & - \tfrac{1}{2}i\theta^\alpha\theta_\alpha\bar{\theta}_\delta\sigma^{\mu\delta\beta}\partial_\mu\psi_\beta + \tfrac{1}{4}\theta^\alpha\theta_\alpha\bar{\theta}_\beta\bar{\theta}^\beta\Box(A - iB).\end{aligned} \tag{14.9}$$

Unlike ϕ_1, the chiral superfield *does* depend on $\bar{\theta}$, but in a manner preordained by the solution (14.7), (14.8) of the chiral constraint.

We note here a few important properties of chiral superfields. We have seen earlier that the product of two superfields is a superfield. Since both $\bar{D}_{\dot{\alpha}}$ and D_α are derivations (i.e., obey Leibniz' rule), the product of two left-chiral (right-chiral) superfields is again a left-chiral (right-chiral) superfield.

Next, consider the superspace integral of a chiral superfield $\int d^4x\, d^2\theta\, d^2\bar{\theta}\phi$. The Berenzin integration over the θs and $\bar{\theta}$s is, as we saw in chapter 9, equivalent to taking the corresponding, θ, or $\bar{\theta}$ derivative. But for a chiral superfield, equations (14.3) and (14.5) imply that either the θ or the $\bar{\theta}$ derivative reduces to a divergence. Hence after performing the Berezin integrals on a superfield, the rest of the superspace integral, namely the space–time integral has an exact divergence as integrand: it is a surface term. The superspace integral of a chiral superfield is thus always a surface term, and as such irrelevant in an action principle. Since we will be dealing with products of chiral superfields, it is worthwhile to note that supersymmetric integrals that are *not* surface terms in ordinary space–time, can be obtained by integrating on the submanifold $\bar{\theta} = 0$ for a left-chiral or

$\theta = 0$ for a right-chiral superfield. Indeed, for a left-chiral superfield ϕ

$$\int d^4x\, d^2\theta \phi(x, \theta, \bar{\theta})|_{\bar{\theta}=0} = \int d^4x\, d^2\theta\, d^2\bar{\theta} \phi(x, \theta, \bar{\theta}) \delta^2(\bar{\theta}) \qquad (14.10)$$

is supersymmetric, and we have a similar formula for right-chiral superfields. We can now proceed to the construction of supersymmetric actions in four space–time dimensions, and then to the study of the remarkable features of the corresponding quantum field theories.

15
The Wess–Zumino model

Given a left-chiral superfield ϕ we want to write down a supersymmetric action. We integrate over superspace $\int d^4x\, d^2\theta\, d^2\bar{\theta}$ (yes, we have to integrate over $\bar{\theta}$, ϕ does depend on $\bar{\theta}$, ϕ_1 of chapter 14 didn't) a lagrangian superfield density $\mathscr{L}$. Based on lower-dimensional experience, we would be tempted to apply some covariant derivatives to ϕ and form $\mathscr{L}$ from these covariant derivatives of ϕ. But, unlike in previous cases, a chiral superfield is constrained, it already contains derivatives of the component fields. Applying any further covariant derivatives, would produce higher derivatives of these component fields, and lead us to an undesirable theory with higher derivatives. So we shall try to make do with the superfield ϕ and its hermitean conjugate (remember the chiral superfield is complex) and take no further covariant derivatives. The simplest term we can contemplate is then $\phi^+\phi$. We could of course also consider ϕ^2 and ϕ^{+2} but these are products of like-chiral superfields and as such chiral superfields themselves. Therefore, their superspace integrals yield uninteresting surface terms as was noted in the last section. But equation (14.10) suggests that we include the terms $\phi^2\delta^2(\bar{\theta}) + \text{h.c.}$ The Grassmann coordinates $\bar{\theta}$ have dimension of $(\text{length})^{1/2}$ so that $\delta^2(\bar{\theta}) = \bar{\theta}_2\bar{\theta}_1$ has dimension of length. Relative to the term $\phi^+\phi$, the $\phi^2\delta^2(\bar{\theta})$ term and its hermitean conjugate must therefore be multiplied with a coefficient m having dimension of mass. Not surprisingly then $\phi^+\phi$ turns out to yield kinetic terms for the component fields whereas $m\phi^2\delta^2(\bar{\theta}) + \text{h.c.}$ is the supersymmetric mass term. One may wonder how the different powers of m arise in the Fermi and Bose mass terms. The mass term in its raw form is overall linear in m, but upon the elimination of the auxiliary fields, the mass terms of the propagating Bose fields acquire the second m factor, much the way the superfield action equation (12.17), linear in g, was able to induce a quartic self-interaction of the bosonic A-field of strength $\frac{1}{2}g^2$. Indeed, an entirely similar construction for an interaction term in four dimensions yields $\int d^4x\, d^2\theta\, d^2\bar{\theta}\phi^3\delta^2(\bar{\theta}) + \text{h.c.}$. Again the $\delta^2(\bar{\theta})$ appears on account of ϕ^3 being left-chiral. So we end up with the Wess–Zumino action (its supersymmetry is proved by a straightforward generalization of the argument between equations (10.7) and (10.10))

$$S_{\mathrm{WZ}} = \int \mathrm{d}^4x\, \mathrm{d}^2\theta\, \mathrm{d}^2\bar{\theta} \{\phi^+\phi + [(m\phi^2 + \tfrac{4}{3}g\phi^3)\delta^2(\bar{\theta}) + \text{h.c.}]\}. \quad (15.1)$$

Inserting here the θ-expansion (14.9) we find that F and G are nonpropagating auxiliary fields. (This could have been realized on dimensional grounds alone. Indeed, as the only Fermi field in the theory, ψ has to propagate, and as such, is expected to have dimension (length)$^{-3/2}$. With θ having dimension (length)$^{1/2}$, we see from equation (14.8) or (14.9) that the fields A, B must have canonical dimension (length)$^{-1}$, whereas the fields F and G must be dimensionless auxiliary fields.) Carrying out the Berezin integrals in (15.1) and eliminating the auxiliary fields F and G via the field equations in the same way as in (12.19), we get the following action in terms of the propagating fields A, B, ψ alone:

$$S_{\mathrm{WZ}} = \int \mathrm{d}^4x[-\tfrac{1}{2}(\partial_\mu A\partial^\mu A + \partial_\mu B\partial^\mu B) - \tfrac{1}{2}\mathrm{i}\bar{\psi}\gamma\partial\psi - \tfrac{1}{2}m^2(A^2 + B^2)$$
$$- \tfrac{1}{2}\mathrm{i}m\bar{\psi}\psi - gmA(A^2 + B^2) - \tfrac{1}{2}g^2(A^2 + B^2)^2 - \mathrm{i}g\bar{\psi}(A - \gamma_5 B)\psi]. \quad (15.2)$$

Here we reverted via equations (5.3) from the Weyl spinor ψ in (14.9) to its Majorana equivalent, which for simplicity is also denoted by ψ in (15.2). The action S_{WZ} defines the Wess–Zumino model. Just as a check, on-shell the two Fermi degrees of freedom of the Weyl spinor ψ are matched by the Bose degrees of freedom of the scalar fields A and B. Off-shell ψ develops two extra degrees of freedom, again matched by those of the auxiliary scalar Bose fields F and G. In chapter 17 we shall look into the quantum field theory built on (15.2).

16

The supersymmetric Maxwell and Yang–Mills theories

Consider two (complex) left-chiral superfields ϕ_1 and ϕ_2. Define

$$\phi = \frac{1}{\sqrt{2}}(\phi_1 + i\phi_2). \tag{16.1}$$

The kinetic action

$$S = \int d^4x\, d^2\theta\, d^2\bar{\theta}\phi^+\phi \tag{16.2}$$

is invariant under the global rephasings

$$\phi \to e^{-ig\lambda}\phi \quad \phi^+ \to e^{ig\lambda}\phi^+ \tag{16.3}$$

where $g\lambda$ is a (superspace-independent) phase angle. So far, everything sounds as in the discussion of ordinary gauge invariance. Indeed, the next step would be to let λ become space-dependent $\lambda(x)$. But that is not possible in a supersymmetric context. For, even if ϕ started out a superfield, $e^{-ig\lambda(x)}\phi$ is not a superfield any longer. This is readily corrected by letting λ acquire not only an x-dependence but also a θ- and θ-dependence so that $\lambda \to \Lambda(x, \theta, \bar{\theta})$ with Λ a *complex* ($\Lambda^+ \neq \Lambda$) chiral superfield (Wess & Zumino 1974*b*). Then

$$\phi \to e^{-ig\Lambda}\phi, \quad \phi^+ \to \phi^+ e^{ig\Lambda^+} \tag{16.4}$$

and

$$\phi^+\phi \to \phi^+\phi \exp[ig(\Lambda^+ - \Lambda)] \neq \phi^+\phi. \tag{16.5}$$

So, the kinetic matter action (16.2) while invariant under the global transformations (16.3), fails to be invariant under the local transformations (16.4). A la Maxwell–Weyl–Yang–Mills we replace

$$\phi^+\phi \Rightarrow \phi^+ e^{gV}\phi \tag{16.6}$$

with V, what we called in chapter 14 a vector superfield, which undergoes the transformation

$$V \to V + ig(\Lambda - \Lambda^+),$$

when the matter fields undergo the transformation (16.4). The chiral superfield Λ has a θ-expansion of the type (14.9). We shall use the notations of equation (14.9) for the component fields of Λ. We can then calculate the change of the vector superfield

$$\delta V = \mathrm{i}g(\Lambda - \Lambda^+). \tag{16.7}$$

Comparing the result with the θ-expansion of the vector field V, equation (14.1), we deduce the changes in the individual component fields of V to be

$$\left.\begin{aligned} &\delta C = B, \quad \delta\chi = \psi, \quad \delta M = F, \quad \delta N = G, \\ &\delta V_\mu = \partial_\mu A, \quad \delta\lambda = 0, \quad \delta D = 0. \end{aligned}\right\} \tag{16.8}$$

The chiral superfield Λ being arbitrary, B, ψ, F, G are arbitrary and can suitably be chosen to cancel out the original C, χ, M and N fields of the vector superfield:

$$B = -C, \quad \psi = -\chi, \quad F = -M, \quad G = -N \tag{16.9}$$

The component fields C, χ, M, N of the vector superfield V are thus recognized as gauge degrees of freedom that can be eliminated by going to the *Wess–Zumino gauge* (16.9). From (16.8) the component field A of the 'phase' superfield Λ is seen to correspond to an ordinary abelian gauge transformation of the abelian gauge field V_μ. In the Wess–Zumino gauge, the θ-expansion of V simplifies considerably

$$V|_{\mathrm{WZ}} = -\theta\sigma_\mu\bar{\theta}V^\mu + \mathrm{i}(\theta\theta)(\overline{\theta\lambda}) - \mathrm{i}(\overline{\theta\theta})(\theta\lambda) + \tfrac{1}{2}(\theta\theta)(\overline{\theta\theta})D. \tag{16.10}$$

$V|_{\mathrm{WZ}}$ has no body, and as such is nilpotent

$$(V|_{\mathrm{WZ}})^n = 0 \quad \text{for } n > 2. \tag{16.11}$$

The supersymmetric, gauge invariant kinetic term $\phi^+ \mathrm{e}^{gV}\phi$ (equation (16.6)), which on account of the exponential is nonpolynomial in most gauges, thus becomes polynomial in a Wess–Zumino gauge, since by (16.11) the Taylor expansion of the exponential then truncates at the quadratic term.

It must be pointed out though, that the Wess–Zumino gauge is *not* supersymmetric. Supersymmetry transformations in general carry a field out of the Wess–Zumino gauge.

To have a full supersymmetric Maxwell theory we still need a gauge invariant superfield containing $F_{\mu\nu} = \partial_\mu V_\nu - \partial_\nu V_\mu$, from which the Maxwell kinetic term and its supersymmetric partners are to be built. By acting on V with sufficiently many covariant derivatives, we can achieve invariance under the transformations (16.7), since these covariant derivatives will annihilate chiral superfields that appear in δV. The superfield

$$W_\alpha = (\bar{D}\bar{D})D_\alpha V \tag{16.12a}$$

and its conjugate

$$\bar{W}_{\dot{\alpha}} = (DD)\bar{D}_{\dot{\alpha}} V \tag{16.12b}$$

have the properties of

(i) being chiral $\bar{D}_{\dot{\alpha}} W_\beta = 0$, $D_\alpha \bar{W}_{\dot{\beta}} = 0$

(ii) being gauge invariant $\delta W_\alpha = 0$, $\delta \bar{W}_{\dot{\alpha}} \sim 0$ under the transformation (16.7)

(iii) obeying the identity $D^\alpha W_\alpha - \bar{D}_{\dot{\alpha}} \bar{W}^{\dot{\alpha}} = 0$.

The action (W_α, $\bar{W}_{\dot{\alpha}}$ are chiral)

$$S_{\text{SUSY MAXWELL}} = \int \mathrm{d}^4x\, \mathrm{d}^2\theta\, \mathrm{d}^2\bar{\theta} \tfrac{1}{32}[W^\alpha W_\alpha \delta^2(\bar{\theta}) + \bar{W}_{\dot{\alpha}} \bar{W}^{\dot{\alpha}} \delta^2(\theta)] \tag{16.13a}$$

once the Berezin integrations are carried out, yields in the Wess–Zumino gauge (16.9), the component field action

$$S_{\text{SUSY MAXWELL WZ}} = \int \mathrm{d}^4x(-\tfrac{1}{4}F_{\mu\nu}F^{\mu\nu} - \tfrac{1}{2}\mathrm{i}\bar{\lambda}\gamma\partial\lambda + \tfrac{1}{2}D^2). \tag{16.13b}$$

We recognize D as a nonpropagating auxiliary field and λ as the supersymmetric partner of the photon field: the *photino* field.

On-shell the propagating photon and photino each have two degrees of freedom, whilst off-shell the full vector multiplet comes into play. The off-shell photino now has doubled the number of its components (it does not obey the Dirac equation) to four and in addition to field χ also weighs in with four components of its own for a grand total of eight Fermi components. On the Bose-side, off-shell, the photon 4-vector has four components with the remaining four being provided by the scalars C, D, M and N.

To summarize, we found for supersymmetric electrodynamics the supersymmetric *and* gauge-invariant action

$$S_{\text{SUSY ED}} = \int \mathrm{d}^4x\mathrm{d}^2\theta\mathrm{d}^2\bar{\theta}\{\phi^+ \mathrm{e}^{gV}\phi + \tfrac{1}{32}[W^\alpha W_\alpha \delta^2(\bar{\theta}) + \bar{W}_{\dot{\alpha}}\bar{W}^{\dot{\alpha}}\delta^2(\theta)]\} \tag{16.14}$$

with the first term providing the coupling to a chiral matter superfield. We refrain here from writing out in detail the component form of this interaction term. Suffice it to say that the photon couples both to charged Fermi matter and to its charged scalar superpartners as expected, whereas the photino couples the Fermi matter fields to their Bose partners, in a matched universal way.

There is one surprise in store. The term

$$S_{\mathrm{FI}} = \kappa \int \mathrm{d}^4x \, \mathrm{d}^2\theta \, \mathrm{d}^2\bar{\theta} V \tag{16.15}$$

with κ a constant is supersymmetric (by the usual argument) *and* gauge invariant (since $\delta V = \mathrm{i}(\Lambda - \Lambda^+)$ is the sum of two chiral superfields each of which superspace integrate to surface terms as shown in chapter 14). Thus such a *Fayet–Iliopoulos term* (Fayet & Iliopoulos 1974) can be added to $S_{\mathrm{SUSY\ ED}}$, with the result that it induces spontaneous supersymmetry breaking. This whole construction can immediately be generalized (Ferrara & Zumino 1974, Salam & Strathdee 1974*b*) to the nonabelian (except for the Fayet–Iliopoulos term which is only possible for an abelian factor of the overall gauge group).

We wish to comment here on one more item. The modern point of view in gauge theory (see for example Eguchi, Gilkey & Hanson 1980) is to view the gauge potentials as components of a connection one-form on a principal fibre bundle, with the field strengths spanning the corresponding curvature two-form. Gauge theories, like general relativity (itself a gauge theory), are thus viewed as geometries. The construction given in this section had virtually nothing to do with the geometry of some 'superfibre bundle.' An alternative approach supersymmetrizing the usual geometric approach to gauge theory is possible and yields results equivalent to those obtained here (see Wess & Bagger 1983). Yet we shall not present this approach because in its present form the supergeometrical formalism lacks the elegance and simplicity of the geometric formalism for ordinary gauge theories. It is to be hoped that this important formalism will be perfected.

17

Supersymmetric quantum field theories and their applications†

Quantum field theories, even of the tame renormalizable type, contain (manageable) divergences in perturbation theory. As a rule, supersymmetric theories also contain such divergences, but less virulent ones than in nonsupersymmetric theories of similar type. Indeed, after years of getting used to the idea that divergences in quantum field theory are simply unavoidable, supersymmetric theories, possibly even some realistic ones, have been produced in which all divergences cancel to all orders in perturbation theory.

We shall briefly state some of the basic results on divergences in supersymmetric quantum field theories, and then heuristically consider some of their potential phenomenological applications.

Consider the Wess–Zumino model described in chapter 15. For this model it has been found that to all orders in perturbation theory, all loop diagrams, evaluated using superspace techniques, involve an integration over both $\int d^2\theta$ and $\int d^2\bar{\theta}$ *without* the presence of either a $\delta^2(\theta)$ or a $\delta^2(\bar{\theta})$. From equation (15.1) this shows that neither the mass terms $m(\phi^2\delta^2(\bar{\theta}) +$ h.c.) nor the interaction terms $(\frac{4}{3}g\phi^3\delta^2(\bar{\theta}) + \text{h.c.})$ get renormalized. All one encounters is a common overall wave-function renormalization of the Bose and Fermi fields

$$\phi_{\text{renormalized}} = Z^{-1/2}\phi \quad \phi = A, B, F, G, \psi. \tag{17.1}$$

There are no mass and coupling constant renormalizations beyond those implied by (17.1). In other words, writing the generic mass and coupling constant renormalizations as

$$m_{\text{renormalized}} = Zm + \delta m, \quad g_{\text{renormalized}} = Z^{3/2}Z'g, \tag{17.2}$$

in the supersymmetric theory we have

$$\delta m = 0, \quad Z' = 1. \tag{17.3}$$

† This brief section obviously cannot do justice to the vast and important topics implied in its title. It is meant to provoke the reader's curiosity rather than satiate it.

What this means is, that if we fix at the classical (tree) level a Bose–Fermi mass-degeneracy and the supersymmetric relations among coupling constants, as in equation (15.2), then this mass-degeneracy and these coupling constant relations will survive in the quantum theory, to all orders in perturbation theory (on the other hand there exist nonperturbative mechanisms of supersymmetry breaking). This result has been invoked as a solution to the pernicious hierarchy problem in grand unified theories (Witten 1981, Dimopoulos & Raby 1981, Dine, Fischler & Srednicki 1981, Sakai 1981).

In grand unification (Langacker 1981) one postulates a gauge theory with a large simple gauge group ($SU(5)$, $O(10)$ are the simplest candidates), which breaks down via a Higgs mechanism to $SU(3)_{\text{color}} \times [SU(2) \times U(1)]_{\text{electroweak}}$ at a typical scale of $\sim 10^{15}$ GeV. The electroweak theory involves a further scale of $\sim 10^2$ GeV, which is set by the ordinary Weinberg–Salam Higgs bosons. One then expects these bosons not to be much heavier than 100 GeV, to insure perturbative unitarity at the electroweak scale. Compared to the grand unification scale, 100 GeV is very light indeed. Moreover scalar bosons (such as the Higgs bosons) are prone to acquire large masses. Interaction with the Higgs bosons of grand unification should render the electroweak Higgs bosons superheavy as well. At the tree level, cancellations can be contrived, but with generically quadratically divergent radiative corrections one still expects squares of the electroweak Higgs masses of the order of some power of the fine structure constant times the square of the grand unification energy scale (gravity may not be harmless in this context either). Were the grand unification to be $N = 1$ supersymmetric, the just discussed no-renormalization theorems would insure that light scalars stay light; for if the supersymmetry were exact, the quadratically divergent radiative corrections due to Fermi loops would exactly cancel those from Bose loops. But supersymmetry cannot be exact, in nature we do not observe mass-degenerate Fermi and Bose particles. If supersymmetry is broken, this would mean that the electroweak Higgs bosons will acquire a mass which will be larger, the larger the supersymmetry breaking, i.e., the larger the difference $(\delta m)^2$ between the $(\text{mass})^2$ of bosons and fermions originally belonging to the same supermultiplet. A Higgs $(\text{mass})^2$ of order $\alpha(\delta m)^2$ is thus induced which, with $\alpha \sim 10^{-2}$, and a Higgs mass of ~ 100 GeV suggests a supersymmetry breaking effect δm of about 1000 GeV. Models implementing this ideology are legion and will not be discussed here. Their most striking prediction (Witten 1981, Dimopoulos & Raby 1981, Dine, Fischler & Srednicki 1981, Sakai 1981, Fayet 1976) is that of supersymmetric partners to all the 'low mass particles'. Thus spin zero sleptons and squarks are to accompany the fermionic leptons and quarks, whereas spin one-half photinos, gluinos, winos, zinos must accompany the photon, gluon, W and Z, etc.... Once supersymmetry breaking effects can occur at around 1000 GeV (the supersymmetry breaking scale itself may be much higher), then the experimental discovery, at least of the lightest supersymmetry partners should be 'around the corner'.

Three comments are in order here. First, the supersymmetry that might be observed 'down' to 1000 GeV is expected to be of the $N = 1$ (and not of the extended $N = 2$, 3, or 4) type. Otherwise, as can be seen from table 5.1 each spin one-half fermion of a supermultiplet would have a partner of the opposite helicity, still in the same supermultiplet, and the theory would be vector-like rather than chiral as is the case in nature (of course there is always the possibility of 'mirror fermions' that complete the vector-like structure at 1000 GeV or so, in which case even $N \geqslant 2$ supersymmetries could be accommodated).

Second, there is the problem of supersymmetry breaking. In the two-dimensional theory of chapter 12, spontaneous supersymmetry breaking was achieved by introducing the linear term $\lambda\phi$ in the superpotential, and adjusting the relative sign of the coefficient λ with respect to that of the cubic term $(g/3)\phi^3$. In four dimensions we already have chiral cubic interaction and quadratic mass terms. These can, without further ado, be supplemented with linear chiral terms $\lambda^i\phi_i\delta^2(\bar{\theta}) + \text{h.c.}$ and it has been shown by O'Raifeartaigh (1975), that with a minimum of three superfields one can conjure up such 'F-like' spontaneous supersymmetry breaking. With an abelian gauge field one has the Fayet–Iliopoulos D-term, equation (16.15), which also leads to spontaneous supersymmetry breaking.

Finally, we mention here that coupling supersymmetric grand unified theories to nonrenormalizable $N = 1$ supergravity at the Planck scale induces nonrenormalizable terms at low energy which affect the scalar potential, as well as other aspects of phenomenology.

We leave the problem of supersymmetric grand unification at this rudimentary level, with the assurance that should a broken $N = 1$ supersymmetry with superpartners around a few TeV be observed in experiments, this topic will then become due for a book in its own right. In the interim we refer to the reviews of Nilles (1984), and of Nanopoulos & Savoy-Navarro (1984), which contain extensive bibliographies.

18
Finite quantum field theories

In the last chapter we saw the customary divergences of quantum field theories getting alleviated by supersymmetry. Over the last few years it has been established that in the case of extended supersymmetry there exist theories for which the divergences are not only alleviated but outright eliminated: the theories are finite, no divergences whatsoever!

The first theory where finiteness was noticed, first to two loops (Jones 1977, Poggio & Pendleton 1977), then to three loops (Grisaru, Roček & Siegel 1980, Avdeev, Tarasov & Vladimirov 1980, Caswell & Zanon 1981), and then established to all orders in perturbation theory (Mandelstam 1983, Howe, Stelle & Townsend 1984, Brink, Lindgren & Nilsson 1983, West 1983, Grisaru & Siegel 1982) was the extended $N = 4$ supersymmetric Yang–Mills theory with arbitrary compact gauge group G, in four space–time dimensions (Brink, Scherk & Schwarz 1977, Gliozzi, Olive & Scherk 1977). The renormalization-group β-function vanishes identically for this theory and to all orders in perturbation theory one finds finite results for all Green's functions. Individual Feynman graphs may diverge but in each order the divergences cancel among the various graphs. This theory may develop spontaneous G-symmetry breaking and monopoles and may be electric–magnetic self-dual (Osborn 1979, Montonen & Olive 1978). Other finite quantum field theories in four space–time dimensions having only $N = 2$ supersymmetry have since been found (Howe, Stelle & West 1983) and even $N = 1$ supersymmetric candidates are being explored (Parkes & West 1984, Jones & Mezincescu 1984, Hamidi & Schwarz 1984), with potential phenomenological aims. There can be no doubt that the discovery of these theories radically changes our ideas about the structure of quantum field theory. The existing proofs of finiteness in four space–time dimensions are quite technical (see, however, chapter 19). We therefore choose to look at finite quantum field theories in two space–time dimensions. Whereas their existence was noticed after the $N = 4$ theory had already been conjectured to be finite, these two-dimensional theories (Curtright & Freedman 1979) were the first for which finiteness was proved to all orders (Alvarez-Gaumé & Freedman 1981). In this case the argument does involve beautiful ideas from the geometry of Kähler manifolds (Lichnerowicz 1955), so we will sketch it here.

In chapter 12 we have written down the action for a free real superfield in two space–time dimensions in the form

$$S_0 = \frac{1}{4}\int d^2x\, d^2\theta \bar{D}\phi D\phi \tag{18.1}$$

or for a set of N free superfields

$$S_{0N} = \frac{1}{4}\int d^2x\, d^2\theta \delta_{ij}\bar{D}\phi^i D\phi^j \tag{18.2}$$

with summation over i, j from 1 to N understood. From this trivial free action we switch to the nontrivial action of a nonlinear σ-model by replacing (Meetz 1969) the Kronecker δ_{ij} by a general superfield-dependent metric $g_{ij}(\phi^k)$

$$S = \frac{1}{4}\int d^2x\, d^2\theta g_{ij}(\phi^k)\bar{D}\phi^i D\phi^j. \tag{18.3}$$

As in chapter 12, we have the θ-expansion of the superfields

$$\phi^i = A^i(x) + \bar{\theta}\psi^i(x) + \tfrac{1}{2}\bar{\theta}\theta F^i(x). \tag{18.4}$$

Using the expression (12.6) for the covariant derivative, performing the θ integrations and eliminating the auxiliary fields, we can rewrite the action (18.3) in component form

$$S = \frac{1}{2}\int d^2x[g_{ij}(A)\partial_\mu A^i \partial^\mu A^j + \mathrm{i} g_{ij}(A)\bar{\psi}^i \gamma^\mu D_\mu \psi^j$$
$$+ \tfrac{1}{6}R_{ijkl}(\bar{\psi}^i\psi^k)(\bar{\psi}^j\psi^l)]. \tag{18.5a}$$

Here

$$D_\mu \psi^i \equiv \partial_\mu \psi^i + \Gamma^i_{jk}\partial_\mu A^j \psi^k \tag{18.5b}$$

and Γ^i_{jk}, R_{ijkl} are the Christoffel symbols and the components of the Riemann curvature tensor of the metric $g_{ij}(A)$ respectively. They make an appearance in the derivative terms of the Taylor expansion of the super-analytic metric $g_{ij}(\phi^k)$ around $\phi^k = A^k$. This expansion truncates on account of the nilpotence of the θ-dependent part of the ϕ^i and therefore only first and second derivatives appear. These derivatives then reassemble in the connection and curvature terms. Had there been higher order terms in the expansion there would have also been covariant derivatives of curvatures.

The action (18.5) has a supersymmetry described by

$$\left.\begin{aligned}\delta A^i &= \bar{\varepsilon}\psi^i \\ \delta\psi^i &= -\mathrm{i}\gamma\partial A^i\varepsilon - \Gamma^i_{jk}(\bar{\varepsilon}\psi^j)\psi^k.\end{aligned}\right\} \tag{18.6}$$

Were it not for the Γ^i_{jk} term, these would be precisely the *on-shell* transformation laws (12.14). This is as it should be, for with $g_{ij}(\phi)=\delta_{ij}$ we recover the free field case of equation (12.14) and in that case $\Gamma^i_{jk}=0$. The Γ^i_{jk} term is needed in the case $g_{ij}\neq\delta_{ij}$ as can be directly checked. Aside from the supersymmetry (18.6), the action (18.5) is also invariant under general coordinate transformations (diffeomorphisms) on the manifold of the fields

$$A^i \to A'^i(A^j), \quad \psi^i \to \psi'^i = \frac{\partial A'^i}{\partial A^j}\psi^j \tag{18.7}$$

(the ψ^i's transform as a contravariant vector).

We now ask under what circumstances does the action (18.5) admit further supersymmetries beyond (18.6). To answer this question we write down the most general form of such extra supersymmetries again in component form.

$$\left.\begin{aligned}\delta A^i &= (f_B)^i{}_j\bar{\varepsilon}^B\psi^j \\ \delta\psi^i &= -\mathrm{i}(h_B)^i{}_j\gamma\partial A^j\varepsilon^B - (S_B)^i{}_{jk}(\bar{\varepsilon}\psi^j)\psi^k - (V_B)^i{}_{jk}(\bar{\varepsilon}^B\gamma^\mu\psi^j)\gamma_\mu\psi^k \\ &\quad - (P_B)^i{}_{jk}(\bar{\varepsilon}\gamma_5\psi^j)\gamma_5\psi^k.\end{aligned}\right\} \tag{18.8}$$

Here $B=1,2,\ldots,N$ labels the supersymmetry, ε^B is a two-component Majorana spinor and the symbols f_B, h_B, S_B, V_B and P_B are functions of the scalar fields A_i. For these supersymmetries to commute with diffeomorphisms (general coordinate transformations) these symbols (the fs, hs, Ss, Vs, and Ps) must all transform as tensors of appropriate rank and type (as indicated by the number of lower case latin indices they carry). For the action S to be invariant under the supersymmetry of parameter ε_B we must have

$$\left.\begin{aligned}g_{ik}(f_B)^k{}_j = g_{jk}(h_B)^k{}_i, \ \nabla_k(f_B)^i{}_j = 0 \\ (f_B)^i{}_m R^m{}_{jkl} = -R^i{}_{mkl}(f_B)^m{}_j \\ (V_B)^i{}_{jk} = (P_B)^i{}_{jk} = 0 \quad (S_B)^i{}_{jk} = \Gamma^i{}_{kl}(f_B)^l{}_j\end{aligned}\right\} \tag{18.9a}$$

The first two of these equations imply the covariant constancy of $(h_B)^i{}_j$

$$\nabla_k(h_B)^i{}_j = 0 \tag{18.9b}$$

We now impose the supersymmetry algebra

$$[Q^A_\alpha, \bar{Q}^{B\beta}] = 2\delta^{AB}(\gamma^\mu P_\mu)_\alpha{}^\beta \tag{18.10}$$

on the transformations (18.8), and obtain using (18.9) the further constraints

$$g_{ij}(f_A)^i{}_k(f_A)^j{}_l = g_{kl} \tag{18.11a}$$

for each A (i.e., no summation over A is implied in (18.11)), and, using matrix notation for the fs,

$$f_A f_B^{-1} + f_B f_A^{-1} = 2\delta_{AB}. \tag{18.11b}$$

One of the supersymmetry generators (18.10), say that for $A = 1$, is the one we already pointed out in equation (18.6), for which

$$(f_1)^i{}_j = \delta^i{}_j. \tag{18.12}$$

Considering equation (18.11b) for $A \neq 1$, $B = 1$, we then find, using (18.12) that

$$f_A = -f_A^{-1} \quad \text{for} \quad A \neq 1 \tag{18.13}$$

or

$$(f_A)^i{}_j(f_A)^j{}_k = -\delta_k{}^i. \tag{18.14}$$

Multiplying equation (18.11a) for $A \neq 1$ by $(f_A)^l{}_m$ and using standard tensor notation $g_{ij}(f_A)^i{}_k = (f_A)_{ik}$, we then find from (18.14)

$$(f_A)_{ij} = -(f_A)_{ji}. \tag{18.15}$$

Finally (18.13) and (18.11b) combine to

$$[f_A, f_B]_+ = 2\delta_{AB}. \tag{18.16}$$

Now to the meaning of all this algebra. A manifold hospitable to a tensor $(f_A)^i{}_j$ obeying (18.14) is said to admit an *almost complex structure*. For, such a tensor acting twice in succession on any tangent vector v^i will effect a multiplication of this vector by -1, just as the imaginary unit i would. The multiplication of the vector v^i by the complex number $a + ib$, $a, b \in \mathbb{R}$ is 'faked' by

$$V^i \to (a\delta^i{}_j + b(f_A)^i{}_j)V^j. \tag{18.17}$$

An almost complex structure obeying (18.11a) is said to be an *almost hermitean structure* on the manifold. But by (18.9a) $(f_A)^i{}_j$ are covariantly constant almost hermitean structures. A $2n$-dimensional manifold with such a covariantly constant almost hermitean structure is called a *Kähler manifold*: It can be covered by charts into $\mathbb{C}^n$, with holomorphic transition functions. Introducing complex local coordinates $z^1, \ldots, z^n$, the line element of such a Kähler manifold is of the form

$$\mathrm{d}s^2 = g_{\alpha\bar{\beta}} \mathrm{d}z^\alpha \mathrm{d}\bar{z}^{\bar{\beta}} \tag{18.18a}$$

where $\bar{z}^{\bar{\alpha}}$ is the complex conjugate of z^α, and no terms of the form $g_{\alpha\beta}\mathrm{d}z^\alpha \mathrm{d}z^\beta$ or $g_{\bar{\alpha}\bar{\beta}}\mathrm{d}\bar{z}^{\bar{\alpha}}\mathrm{d}\bar{z}^{\bar{\beta}}$ appear. The covariant constancy of the almost hermitean structure translates into

$$\frac{\partial}{\partial z^\gamma} g_{\alpha\bar{\beta}} - \frac{\partial}{\partial z^\alpha} g_{\gamma\bar{\beta}} = 0 \tag{18.18b}$$

once complex coordinates are introduced. We thus recognize the two-dimensional version of *Zumino*'s *theorem* (Zumino 1979):

- There exists precisely one supersymmetry beyond the supersymmetry (18.6) if the manifold of the scalar fields is Kähler.
  Alternatively: a two-dimensional σ-model has $N = 2$ supersymmetry if the scalar field manifold is Kähler.

It is readily checked that any two-dimensional σ-model which admits $N = 3$ supersymmetry also admits $N = 4$ supersymmetry, and that in this case the scalar field manifold must be *hyperkähler*, (Calabi 1979), i.e., it has a quaternionic structure in the tangent space. This quaternionic structure uses the $(f_1)^i{}_j = \delta^i{}_j$ of equation (18.6) for the real part, and the three almost complex structures f_2, f_3, f_4 as 'faking' the three imaginary units of quaternions à la (18.17). It is also readily established that there exist no two-dimensional σ-models with $N > 4$ supersymmetry.

After all this, admittedly pretty, preparation, we now finally reach the crucial problem, that of establishing the ultraviolet finiteness of the theory (18.5) if it admits $N = 4$ supersymmetry, i.e., if the manifold of the scalars is hyperkähler. Briefly, the argument goes as follows. Nonsupersymmetric σ-models have been extensively studied (Friedan 1980), and extending these results to the supersymmetric case one finds that the ultraviolet counterterms are of the type

$$\delta S = \frac{1}{4\mathrm{i}} \int \mathrm{d}^2 x \mathrm{d}^2 \theta \, T_{ij}(\phi) \bar{D}\phi^i D\phi^j$$

with the tensor T_{ij} an algebraic expression in terms of the curvature tensor constructed from the metric $g_{ij}(\phi)$ and of the covariant derivatives of this curvature tensor. The l-loop contribution $T_{ij}^{(l)}$ to T_{ij} is found to have conformal weight $l - 1$, i.e., under the conformal transformation $g_{ij} \to \Lambda^{-1} g_{ij}$ with $\Lambda = \text{constant}$, $T_{ij}^{(l)} \to \Lambda^{l-1} T_{ij}^{(l)}$. These counterterms are 'on-shell', there being also 'off-shell' counterterms that vanish upon the use of the field equations. These off-shell counterterms are of no importance, as they can always be compensated by field redefinitions.

The effective unrenormalized metric is $g_{ij} + T_{ij}$. Now g_{ij} is hyperkähler by assumption, and if $N = 4$ supersymmetry is to survive (as can be shown it must; no anomalies) $g_{ij} + T_{ij}$ must also be hyperkähler. All hyperkähler manifolds are Ricci-flat (Lichnerowicz 1955). Thus both the Ricci tensors calculated from g_{ij} and from $g_{ij} + T_{ij}$ must vanish. This can be shown to require T_{ij} to be a zero mode of the Lichnerowicz laplacian Δ_L. A particular case of hyperkähler manifolds are the four-real-dimensional gravitational ALE-instantons (Gibbons & Hawking 1978, Hitchin 1979) for which all zero modes of Δ_L are known to have conformal weight -1. But at $l > 0$ loops we saw $T_{ij}^{(l)}$ having conformal weight $l - 1 > -1$. Thus no ultraviolet counterterms can arise and these $N = 4$ supersymmetric σ-models in two dimensions are on-shell ultraviolet finite to all orders in perturbation theory!

19
The supercurrent and anomaly supermultiplets

When the fields that appear in the lagrangian of a classical theory undergo the transformations dictated by an ordinary Lie algebra, the change of the action itself, according to the celebrated theorem of Emmy Noether, can be cast into the form of a space–time integral of the divergence of some currents. In particular, when the action is invariant under the effect of the Lie algebra these Noether currents (their number equal to the Lie algebra's dimension) are conserved. Not surprisingly, this theorem admits a straightforward generalization to the case of Lie superalgebras, i.e., to the supersymmetric case. Along with the vectorial and tensorial Bose currents, there will exist spinorial Fermi currents in this case (see chapter 1). In both ordinary and supersymmetric quantum theories, *anomalies* can break the classical conservation laws.

We shall discuss here the classical conservation laws and their quantum demise via anomalies in a supersymmetric theory. As we shall see, the existence of a supermultiplet containing both the energy–momentum tensor and the chiral current at the classical level can lead to some paradoxical results in the quantum theory, unless sufficient care is exercised.

Consider a field theory which classically is both $N = 1$ Poincaré supersymmetric and conformally invariant, i.e., invariant under the full $N = 1$ superconformal algebra $su(2,2|1)$ (see chapter 4).

Now let us look at the conserved currents. As in the non-supersymmetric case these fall into two classes:

(A) the currents that depend only on the fields and their gradients;
(B) the currents that depend explicitly on the space–time coordinates.

Of the Bose currents, the energy–momentum tensor $\theta_{\mu\nu}$ and the chiral current j^5_μ are of type (A), whereas the angular momentum–Lorentz boost tensor $M_{\mu\nu\rho}$, the dilatation current j^D_μ and the conformal current $K_{\mu\nu}$ are of type (B) (e.g., $j^D_\mu = x^\nu\theta_{\mu\nu}$, $M_{\mu\nu\rho} = x_\mu\theta_{\nu\rho} - x_\nu\theta_{\mu\rho}$, etc...) Not surprisingly, of the Fermi supercurrents, the supertranslation current S^α_μ (the 'square root' of $\theta_{\mu\nu}$) is of type (A) and the super-conformal current T^α_μ (the 'square root' of $K_{\mu\nu}$) is of type (B). The $N = 1$ Poincaré superalgebra is a subsuperalgebra of $su(2,2|1)$ and as such, the just listed $su(2,2|1)$ currents transform reducibly under the Poincaré superalgebra. In particular, all type (A) currents form a supermultiplet as noted by Ferrara & Zumino (1975). Thus the supersymmetric partners of the energy–momentum tensor are the Poincaré supercurrent (corresponding to supertranslations) and the chiral current.

Classically all $su(2,2|1)$ currents are obviously conserved for a superconformally invariant theory. At the quantum level super-Poincaré invariance is maintained, whilst the $su(2,2|1)$ invariance is broken by anomalies. Even without supersymmetry one encounters the *a priori* unrelated chiral and conformal (or trace) anomalies. The former breaks the conservation of the chiral current $\partial_\mu j^{5\mu} \neq 0$, the latter the conservation of the dilatation and conformal currents $\partial_\mu j^{D\mu} \neq 0$ $\partial_\mu K^\mu_\alpha \neq 0$. For the dilatation current $j^D_\mu = x^\nu \theta_{\mu\nu}$, on account of the conservation of $\theta_{\mu\nu}$ even at the quantum level, a non-vanishing divergence amounts to a non-vanishing trace θ of $\theta_{\mu\nu}$. So does the nonconservation of $K_{\mu\nu}$. Thus, because of the chiral anomaly, $\partial_\mu j^{5\mu}$ does not vanish, and because of the conformal anomaly, θ does not vanish. In the supersymmetric context, j^5_μ and $\theta_{\mu\nu}$ belong to the same supermultiplet. One then expects $\partial_\mu j^{5\mu}$ and θ to also belong to the same supermultiplet, with the Fermi sector being provided by the divergence $\partial_\mu T^{\mu\alpha}$ of the superconformal (not supertranslational!) current T^α_μ. Just as the $K_{\mu\nu}$s are expressed in terms of $\theta_{\mu\nu}$ and of the coordinates x^ρ, so that $\partial_\mu K^\mu_\nu \neq 0$ implies $\theta \neq 0$, so $T_{\mu\alpha}$ is expressed in terms of the supertranslational current $S_{\mu\alpha}$, of the coordinates x^ρ, and of γ matrices

$$T^\alpha_\mu = (\gamma^\nu x_\nu)^\alpha_\beta S^\beta_\mu \tag{19.1}$$

with the result that $\partial_\mu T^{\mu\alpha} \neq 0$ implies an anomaly in the 'γ-trace' (not the divergence!) of the supertranslational current (or supercurrent) S^α_μ

$$(\gamma^\mu S_\mu)^\alpha \neq 0. \tag{19.2}$$

This is the superconformal anomaly (Curtright 1977, Abbott, Grisaru & Schnitzer 1977). So one expects $\partial_\mu j^{5\mu}, \theta$ and $\gamma^\mu S^\alpha_\mu$ to form a chiral supermultiplet. The previously independent chiral and trace anomalies are now connected and further tied to the superconformal anomaly. Breaking of conformal and chiral invariance always were believed to go hand-in-hand (Carruthers 1971), and here this really happens. Or does it? The famous Adler–Bardeen theorem (Adler & Bardeen 1969) states that the chiral anomaly is strictly a one-loop phenomenon, whereas, on the other hand, the trace anomaly is proportional to the renormalization group β-function and as such acquires contributions from two and more loops. How can these anomalies possibly have anything to do with each other? This is the paradox alluded to at the beginning of this section.

Now to the resolution of this paradox. At the classical level all the $su(2,2|1)$ currents are conserved. At the quantum level the corresponding operators are only rendered meaningful once one also prescribes a regularization procedure (Clark, Piguet & Sibold 1978, Grisaru & West 1985). One given classical current can thus generate *more* than one renormalized quantum current depending on the regularization procedure. Thus we expect to find at the quantum level both a chiral current $j^5_{\mu\mathrm{AB}}$ obeying the Adler–Bardeen theorem but not related by supersymmetry to the energy–momentum tensor, and a chiral current $j^5_{\mu\mathrm{SUSY}}$ violating this theorem but representing the supersymmetric partner of $\theta_{\mu\nu}$. Classically, $j^5_{\mu\mathrm{AB}}$ is indistinguishable from $j^5_{\mu\mathrm{SUSY}}$ but quantum mechanically they are represented by different operators.

Let us illustrate all this with a nontrivial example. Consider the $N=1$ supersymmetric Yang–Mills theory. It is classically conformally invariant and therefore exhibits superconformal $su(2,2|1)$ invariance. Just as in chapter 16, where we dealt with the supersymmetric Maxwell theory, the supersymmetric Yang–Mills theory contains in the Wess–Zumino gauge (after the elimination of the auxiliary fields) a massless vector field v^i_μ and a massless Majorana spinor λ^i_α both in the adjoint representation of the gauge group $G(i=1,\ldots,\dim G)$. The lagrangian is that of Yang–Mills for v^i_μ, with a minimal coupling of the gauge field to the spinor fields. At the classical level, the chiral current $j^5_\mu = \bar{\lambda}^i\gamma_\mu\gamma_5\lambda^i$ is conserved, the energy–momentum tensor $\theta_{\mu\nu}$ is traceless and the supercurrent S^α_μ obeys $\gamma^\mu S_\mu = 0$. The use of a supersymmetric subtraction scheme then defines a quantum operator $j^5_{\mu\mathrm{SUSY}}$, which will share a supermultiplet with the energy–momentum tensor. Calculating its divergence one finds (Grisaru & West 1985)

$$\partial_\mu j^{5\mu}_{\mathrm{SUSY}} = -\frac{1}{3}\frac{\beta(g)}{g}[\tfrac{1}{2}(\varepsilon^{\mu\nu\rho\sigma}F^i_{\mu\nu}F^i_{\rho\sigma})_{\mathrm{SUSY}} - \partial_\mu\bar{\lambda}^i\gamma^\mu\gamma_5\lambda^i] \tag{19.3}$$

with $\beta(g)$ the renormalization group β-function of the supersymmetric Yang–Mills theory under investigation. Here the $^*FF$ term is also supersymmetrically regularized. The bracketed expression on the right-hand side of equation (19.3) is the G-like component (in the notation (14.9)) of a chiral superfield whose F-like component is precisely the supersymmetric Yang–Mills lagrangian, and indeed $-\frac{4}{3}(\beta(g)/g)\mathscr{L}$ is what appears in the trace anomaly. The Fermi spinor component of this same superfield then yields the superconformal anomaly. As expected, with a supersymmetric subtraction scheme we find a chiral superfield that has the three anomalies as three of its component fields.

But... all this appears to run in the face of the Adler–Bardeen theorem. This theorem claims the existence of a gauge invariance maintaining 'AB subtraction scheme' which yields a current $j^5_{\mu\mathrm{AB}}$ for which

$$\partial_\mu j^{5\mu}_{\mathrm{AB}} = \frac{g^2}{16\pi^2} C_2(G) \tfrac{1}{2} (\varepsilon_{\mu\nu\rho\sigma} F^{i\mu\nu} F^{i\rho\sigma})_{\mathrm{AB}} \tag{19.4}$$

with $C_2(G)$ the value of the quadratic Casimir operator for the gauge group G. The one-loop (cubic) term $\beta_1(g)$ of $\beta(g)$ is given by

$$\beta_1(g) = -\frac{3g^3}{16\pi^2} C_2(G) \tag{19.5}$$

so that

$$\partial_\mu j^{5\mu}_{\mathrm{AB}} = -\frac{\beta_1(g)}{3g} \tfrac{1}{2} (\varepsilon_{\mu\nu\rho\sigma} F^{i\mu\nu} F^{i\rho\sigma})_{\mathrm{AB}} \tag{19.6}$$

and the Adler–Bardeen anomaly (19.6) differs from the supersymmetric one on two counts: (i) it does not contain the second term in the bracket of (19.3) and (ii) it is multiplied by the one-loop β-function β_1, rather than by the full β-function as in (19.3). It is *not* the superpartner of the trace and superconformal anomalies. Remarkably there exists a *non-supersymmetric* subtraction procedure which produces the Adler–Bardeen theorem obeying current. Without going into all the details (Grisaru & West 1985) this subtraction procedure removes the $\partial_\mu \bar{\lambda}^i \gamma^\mu \gamma^5 \lambda^i$ piece at order g^2, as required by the Adler–Bardeen theorem, but then at order g^4 it produces a new $*FF$ term with coefficient $(\beta_2/3g) + 2(\beta_1/3g)^2$ (where β_2 is the two-loop term in the β-function). The Adler–Bardeen theorem forbids such an extra $*FF$ term, thus determining β_2 in terms of β_1. Iterating this reasoning one determines β_3 in terms of β_2 and β_1 i.e., in terms of β_1 and so on. In the end one determines all higher loop terms β_n in the β-function (β_n for $n \geqslant 3$ are subtraction scheme dependent!) leading to the closed Jones formula (Jones 1983) for this function

$$\beta(g) = \frac{\beta_1(g)}{1 + \dfrac{2\beta_1(g)}{3g}}. \tag{19.7}$$

We have thus removed the paradox. The solution is that the same classical chiral current $\bar{\lambda}^i \gamma_\mu \gamma_5 \lambda^i$ has more than one quantum operator counterpart. There is $j^5_{\mu\mathrm{SUSY}}$ whose divergence completes the anomaly supermultiplet and $j^5_{\mu AB}$ which obeys the Adler–Bardeen theorem. This is all very nice but it raises a phenomenological question (Curtright 1984). Suppose the grand unified theory exists and is supersymmetric to start with. Can we then observe matrix elements of both $j^5_{\mu AB}$ and $j^5_{\mu SUSY}$, or only of one, and in that case which one? Does the chiral current whose anomaly accounts for the decay $\eta^0 \to 2\gamma$ necessarily obey the Adler–Bardeen theorem as is always assumed, without reference to an underlying supersymmetric gauge theory?

The arguments presented in this section have one further interesting application. The validity of the Adler–Bardeen theorem allowed us to calculate the full β-function from its one-loop expression. One can repeat this argument for extended $N \geqslant 2$ supersymmetric Yang–Mills theories coupled to matter with the result

$$\beta(g) = \beta_1(g) \tag{19.8}$$

instead of the Jones formula (19.7). In $N=2$ gauge theories, $N=2$ matter supermultiplets can be introduced in a variety of ways which will rig up the vanishing of $\beta_1(g)$ and therefore, by (19.8), of $\beta(g)$. In particular, the $N=4$ supersymmetric Yang–Mills theories, which can of course be viewed as $N=2$ theories with $N=2$ matter-supermultiplets completing the $N=4$ particle contents, such a rigging of $\beta_1(g)$ is known to occur. But there are many more such theories (Howe, Stelle & West 1983). If we accept the Adler–Bardeen theorem, we then possess a proof of the vanishing of the β-function in all these theories. Power counting arguments (West 1983, Grisaru & Siegel 1982, Howe, Stelle & Townsend 1984) show that the only divergences in theories with $N \geqslant 2$ are those that lead to $\beta_1(g) \neq 0$. This rigging of $\beta_1(g) = 0$ in all these theories then implies that they all are ultraviolet finite quantum field theories.

On the other hand, one disposes of direct proofs of the finiteness of these theories which do not assume the validity of the Adler–Bardeen theorem. As we already mentioned in chapter 18, it has been speculated that there may even exist finite $N=1$ supersymmetric theories.

Part III

Supergravities: locally supersymmetric theories

20
The problem of gauging supersymmetry

The fundamental symmetries exhibited by the laws of nature are local rather than global, i.e. they are gauged. This is certainly true, say, of $SU(3)_{\text{color}}$ and $[SU(2) \times U(1)]_{\text{electroweak}}$. If the laws of nature turn out to exhibit some form of supersymmetry, it is then natural to ask whether supersymmetry can also be gauged. As we saw, all ($N = 1$ and $N > 1$) supersymmetries in four dimensions contain the Poincaré (or de-Sitter) algebra as a subalgebra. Gauging supersymmetry thus implies gauging the Poincaré or de-Sitter algebras. But such a gauge theory of the Poincaré or de-Sitter algebras is the Einstein theory of gravity (Kibble 1961). When gauging supersymmetry, gravity then *must* be included. This gives us an idea about the gauge fields to be expected when supersymmetry is gauged. Gravity is described by a massless spin two boson. Then at the very least, for local $N = 1$ supersymmetry we expect this spin two graviton to acquire a supersymmetric partner. Purely on representation–theoretic grounds (see chapter 5) this partner must be a massless fermion of spin three-halves or five-halves. The supersymmetry charges Q_α span a (spin one-half) Majorana spinor. The corresponding gauge field having one vector index beyond that of the charges, will describe spin three-halves, not five-halves. So the $N = 1$ *supergravity* multiplet contains one massless spin two boson, the graviton, and one Majorana spin three-halves fermion (Volkov & Akulov 1973) the *gravitino*. Theories involving interacting spin three-halves fields are notorious in that they lead to all kinds of inconsistencies including acausal propagation at the classical level (Velo & Zwanziger 1969). It is remarkable that local supersymmetry removes these problems, as we shall see in chapter 21. The problem is now to construct a super-symmetrization of the Einstein–Hilbert lagrangian and to couple the resulting *supergravity* theory to $N = 1$ supersymmetric matter.

The Einstein–Hilbert action has an elegant, simple and well-known interpretation in the context of Riemannian geometry. One's first impulse in constructing the supergravity lagrangian would then lead to a super-Riemannian formulation (Arnowitt & Nath 1976). Alas, its mathematical appeal notwithstanding, such a formulation is of no physical interest. To understand why this is so, we need but recall that in a small enough neighborhood of the world manifold, or more precisely in the tangent space at any point of this world manifold, we have an $SO(3,1)$ invariant Minkowski metric and the laws of special relativity apply. In the $N = 1$ supersymmetric context we deal with a (4,4)-supermanifold and the straightforward extension of the Riemannian idea is to require that in the tangent space an $OSp(4|4)$ metric prevails. Thus locally the laws are $OSp(4|4)$ invariant. But we only expect $OSp(1|4)$ (more precisely its Poincaré contraction) to be operative as in the flat superspace of global supersymmetry (for $N > 1$, again, one gets $OSp(4|4N)$ instead of $OSp(N|4)$). Were one to go ahead anyway, this discrepancy would come to take its revenge, in that the supersymmetrized Hilbert–Einstein action would allow a flat, torsion-free $OSp(4|4)$ invariant 'vacuum'-solution, but would admit *no* Poincaré or de-Sitter supersymmetric vacuum. The superspace of global supersymmetry is *not* a solution of the super-Riemannian field equations.

Another way of seeing the same thing is to recall that ordinary Minkowski space is flat, and in addition to being Riemannian, is also torsionless. By contrast, on account of the expression for the $[D_\alpha, \bar{D}_{\dot{\alpha}}]$ bracket of equation (14.4), the flat superspace of global supersymmetry is *not* torsionless, as required by super-Riemannian geometry (nor can this torsion be eliminated in the standard way).

Thus, constructing supergravity in a supermanifold formalism is not a straightforward geometric task. To be sure, such a construction is possible, but in its present form sufficiently complicated to foster the belief that the final word on superspace–supergravity has not yet been spoken. The Christoffel zero-torsion constraints, are replaced by a set of torsion constraints, of a much less intuitive nature (Wess & Zumino 1977). The major open problem is to understand these constraints in a simple geometric and/or physical way. Given this situation, I will not present superspace–supergravity here. Existing monographs (Wess & Bagger 1983, Gates, Grisaru, Roček & Siegel 1983, van Nieuwenhuizen 1981) cover this topic anyway. Rather, I will use a component approach in which supersymmetry is not manifest and has to be checked at each step. Yet such a component formalism can be given a rather elegant form in four space–time dimensions in the unextended $N = 1$ case. The $N = 1$ supergravity action was first constructed using a component form using a so-called second order formalism (Freedman, van Nieuwenhuizen & Ferrara 1976) and then treated in an elegant first order formalism (Deser & Zumino 1976). We shall follow a variant (MacDowell & Mansouri 1977, Chamseddine & West 1977) that for $N = 1$ most closely parallels the nonsupersymmetric ('$N = 0$') Einstein case.

21
Einstein gravity as a gauge theory

We present here a construction (MacDowell & Mansouri 1977) of ordinary ('$N=0$') Einstein gravity with or without a cosmological term in four-dimensional space–time, to set the stage for an entirely parallel construction for the $N=1$ supersymmetric case. This construction has the added advantage of revealing certain assumptions implied in Einstein's theory, and the possibilities connected with their relaxation.

Viewed as a gauge theory, Einstein's theory is ubiquitous. Whether the gauge group be the Poincaré or the de-Sitter group, we expect a set of gauge fields ω_μ^{ab} for the Lorentz group and a further set e_μ^a for the translations, or the P_a transformations of the de-Sitter group. Everybody knows though, that Einstein's theory contains but one spin two field, originally chosen by Einstein as $g_{\mu\nu}=e_\mu^a e_\nu^b \eta_{ab}$ ($\eta_{ab}=$ Minkowski metric). What happened to the ω_μ^{ab}? The field equations obtained from the Hilbert–Einstein action by varying the ω_μ^{ab} are algebraic in the ω_μ^{ab} and obviously are in the right number to make a solution possible, thus permitting us to express the ω_μ^{ab} in terms of the e_μ^as. The ω_μ^{ab} do not propagate, they are 'composite' fields. We shall see all this happening but mention it at this early stage to allay all uneasiness on the reader's part.

We start from the four-dimensional de-Sitter algebra $sp(4)=so(3,2)$. Technically, this is the anti-de-Sitter algebra, as opposed to the alternative $so(4,1)$ de-Sitter algebra, but only $so(3,2)$ can be supersymmetrized as noted in chapter 4, so we stick to it. Let G_A $A=1,\ldots,10$, be a basis of $sp(4)$, thus defining the structure constants f^C_{AB} via

$$[G_A, G_B]=f^C_{AB}G_C. \tag{21.1}$$

We envision space–time as a four-dimensional manifold M. At each point of M we have a copy of $SO(3,2)$ (a fibre, in fibre-bundle terminology) and we introduce the gauge potentials (the connection) $h_\mu^A(x)$, $A=1,\ldots,10$, $\mu=1,\ldots,4$. Here x are local coordinates on M. From these potentials h_μ^A we calculate the field-strengths (curvature components)

$$\bar{R}^A_{\mu\nu}=\partial_\mu h_\nu^A-\partial_\nu h_\mu^A+f^A_{BC}h_\mu^B h_\nu^C. \tag{21.2}$$

Under an infinitesimal local (x-dependent) $sp(4)$ transformation $\varepsilon^A(x)G_A$, the connection transforms as

$$\delta_\varepsilon h_\mu^A = D_\mu \varepsilon^A \equiv \partial_\mu \varepsilon^A + f^A_{BC} h_\mu^B \varepsilon^C \tag{21.3}$$

whereas the curvatures transform homogeneously

$$\delta_\varepsilon \bar{R}^A_{\mu\nu} = f^A_{BC} \varepsilon^B \bar{R}^C_{\mu\nu}. \tag{21.4}$$

The algebraic Jacobi identity entails the so-called Bianchi identities

$$D_\lambda \bar{R}^A_{\mu\nu} + D_\mu \bar{R}^A_{\nu\lambda} + D_\nu \bar{R}^A_{\lambda\mu} = 0. \tag{21.5}$$

We now wish to write down the action S as an integral over the four-manifold M, which should be invariant under general coordinate transformations on M. We further require the action to depend on the $\bar{R}$s only, and not on a possible Riemannian metric on M. Given that the curvatures $\bar{R}^A_{\mu\nu}$ are the components of a two-form $\bar{R}^A$, this means

$$S(Q) = \int_M \bar{R}^A \wedge \bar{R}^B Q_{AB} \tag{21.6a}$$

where Q_{AB} are constants, yet to be chosen. In ordinary tensor notation, without any reference to differential forms, the ansatz (21.6*a*) amounts to requiring the integrand of S to be quadratic in the curvature components

$$S(Q) = \int d^4x \varepsilon^{\mu\nu\rho\sigma} \bar{R}^A_{\mu\nu} \bar{R}^B_{\rho\sigma} Q_{AB}. \tag{21.6b}$$

We reemphasize that this action is independent of any specific metric $g_{\mu\nu}$ on M. Indeed, general covariance dictates the combinations $d^4x(|g|)^{1/2}$ ($g \equiv \det g_{\mu\nu}$) and $\varepsilon^{\mu\nu\rho\sigma}/(|g|)^{1/2}$ so that in the end, the $g_{\mu\nu}$-dependence via $(|g|)^{1/2}$ cancels.

For later use we record here the response of the action to variations of the connection components h^A_μ. First, when the potentials undergo the local infinitesimal $sp(4)$ transformations (21.3), we see from (21.4) that the variation of the action is

$$\delta_\varepsilon S(Q) = \int d^4x \varepsilon^{\mu\nu\rho\sigma} 2\varepsilon^C f^A_{CD} R^D_{\mu\nu} R^B_{\rho\sigma} Q_{AB}. \tag{21.7}$$

Under general variations δh^A_μ of the connection coefficients we find

$$\delta S(Q) = \int d^4x \delta h^A_\mu [4 h^B_\nu \bar{R}^C_{\rho\sigma} (f^D_{BC} Q_{AD} - f^D_{AB} Q_{DC}) \varepsilon^{\mu\nu\rho\sigma}]. \tag{21.8}$$

We are now in a position to make a meaningful choice of the, so far arbitrary, coefficients Q_{AB} which determine the action $S(Q)$. The equation (21.8) has an important consequence concerning this choice. A 'natural' choice for the numerical coefficients Q_{AB} would seem to be the Killing metric

$$Q_{AB} = G_{AB} = f^{D}_{AC} f^{C}_{BD} \tag{21.9}$$

of the de-Sitter algebra. But, this being a simple algebra, we have

$$f^{D}_{BC} G_{DA} - f^{D}_{AB} G_{DC} = 0 \tag{21.10}$$

so that by (21.8) the variation of the action $S(g)$ vanishes automatically. The lagrangian is then an exact divergence; the action a topological invariant, a surface term. Less trivial choices for Q_{AB} are therefore needed. To effect them we require

(i) the invariance of $S(Q)$ under local Lorentz transformations, i.e. under transformations of type (21.3), (21.7) with ε restricted to the $SO(3,1)$ Lorentz subgroup of $SO(3,2)$;
(ii) the invariance of $S(Q)$ under space inversions.

According to requirement (1) the Q_{AB}s must be constructed only from invariant tensors of the Lorentz group. To see what this means in detail let us introduce the following notation. The capital indices A, B have been assumed ten-valued. Let us achieve this ten-valuedness of say A, by the juxtaposition of a four-valued index $a(a = 0, 1, 2, 3)$ and of a six-valued antisymmetric pair $[a'a'']$ of four-valued indices a', a''

$$\begin{aligned} A = 1, 2, \ldots, 10 &\equiv (a = 0, 1, 2, 3) \oplus ([a'a''] \\ &= [0\,1], [0\,2], [0\,3], [1\,2], [1\,3], [2\,3]). \end{aligned} \tag{21.11}$$

Then the most general Lorentz invariant form of Q^{AB} is

$$Q_{AB} = \begin{cases} \alpha\varepsilon_{aa'bb'} & \text{for } A = [aa'], B = [bb'] \\ \beta\eta_{ab} & \text{for } A = a, B = b \\ 0 & \text{otherwise} \end{cases} \tag{21.12}$$

with α, β two real parameters. In view of the appearance of $\varepsilon^{\mu\nu\rho\sigma}$ in (21.6b), the requirement (2) of space inversion invariance fixes

$$\beta = 0 \tag{21.13}$$

in the expression (21.12). We will therefore concentrate on

$$Q^{(\varepsilon)}_{AB} = \begin{cases} \varepsilon_{aa'bb'} & \text{for } A = [aa'], B = [bb'] \\ 0 & \text{otherwise} \end{cases} \tag{21.14}$$

where, without any loss of generality we have set $\alpha = 1$.

One may wonder now whether, with the choice (21.14), the action becomes invariant under all local de-Sitter transformations, not just the Lorentz transformations, directly imposed in (21.14). Actually this is the case as we shall see below.

We now write out the action $S(Q^{(\varepsilon)})$

$$S(Q^{(\varepsilon)}) = \int d^4x \varepsilon^{\mu\nu\rho\sigma} \varepsilon_{abcd} \bar{R}^{[ab]}_{\mu\nu} \bar{R}^{[cd]}_{\rho\sigma}. \tag{21.15}$$

Here we used the notation (21.11). The curvatures $\bar{R}^{[ab]}_{\mu\nu}$ calculated according to (21.2) with the full de-Sitter $sp(4)$ structure constants f^C_{AB}, ae not the same as the usual curvatures, $R^{[ab]}_{\mu\nu}$, encountered in Einstein's theory. These $R^{[ab]}_{\mu\nu}$s are calculated with the Lorentz structure constants, not the de-Sitter ones. Given that the $sp(4)$ structure constants $f^{[de]}_{a[bc]}$ automatically vanish (in the notation of chapter 4, $[M,P] \sim P$, equation (4.1)), it then follows that

$$\bar{R}^{[cd]}_{\mu\nu} = R^{cd}_{\mu\nu} + h^e_\mu h^f_\nu f^{[cd]}_{ef} \tag{21.16}$$

where, to conform to standard notation we have dropped the anti-symmetrizing bracket on the usual (Lorentz) curvature $R^{cd}_{\mu\nu} \equiv R^{[cd]}_{\mu\nu}$. Following a rescaling of the type (4.2), we can express

$$f^{[ab]}_{cd} = -4\lambda^2(\delta^a_c \delta^b_d - \delta^b_c \delta^a_d). \tag{21.17}$$

Inserting (21.16) and (21.17) into (21.15) we find

$$\begin{aligned} S(Q^{(\varepsilon)} = & \int d^4x \varepsilon^{\mu\nu\rho\sigma} \varepsilon_{abcd} R^{ab}_{\mu\nu} R^{cd}_{\rho\sigma} - 16\lambda^2 \int d^4x \varepsilon^{\mu\nu\rho\sigma} \varepsilon_{abcd} R^{ab}_{\mu\nu} h^c_\rho h^d_\sigma \\ & + 64\lambda^4 \int d^4x \varepsilon^{\mu\nu\rho\sigma} \varepsilon_{abcd} h^a_\mu h^b_\nu h^c_\rho h^d_\sigma. \end{aligned} \tag{21.18}$$

We notice that the first term on the right-hand side, now with ordinary Lorentz curvatures, is precisely the topological Euler invariant (related to the number of 'handles' of the manifold M). It is a surface term, its variation vanishes identically and it can be discarded at the classical level. This is interesting, since we know the Hilbert–Einstein lagrangian to be linear in curvatures. Yet our starting point (21.15) was quadratic in de-Sitter curvatures ($\bar{R}$s), therefore in Lorentz curvatures (Rs). Fortunately the quadratic term, as we just found, is an exact divergence and what survives are the linear 'crossed' terms $\sim \lambda^2$ and the Lorentz curvature independent terms $\sim \lambda^4$. We now proceed to carry out the algebra in (21.18). To this end we make the following notational changes. We relabel $h^a_\mu \equiv e^a_\mu$ and

$h_\mu^{[ab]} \equiv \omega_\mu^{ab}$ so that with (21.2), (21.16) and

$$f^{[ef]}_{[ab][cd]} = \tfrac{1}{4}[\eta_{bc}\delta^e_a\delta^f_d - \eta_{ac}\delta^e_b\delta^f_d + \eta_{ad}\delta^e_b\delta^f_c - \eta_{bd}\delta^e_a\delta^f_c] - e \leftrightarrow f \quad (21.19)$$

we find the usual expressions

$$R^{ab}_{\mu\nu} = \partial_\mu\omega^{ab}_\nu - \partial_\nu\omega^{ab}_\mu + (\omega^{ac}_\mu\omega^{db}_\nu - \omega^{ac}_\nu\omega^{db}_\mu)\eta_{cd} \quad (21.20)$$

e^a_μ being the usual *vierbein* and ω^{ab}_μ the *spin connection* (Ricci rotation coefficients). We further assume (here for the first time!) that the vierbein has an inverse e^μ_a defined by

$$e^\mu_a e^a_\nu = \delta^\mu_\nu \quad e^\mu_a e^b_\mu = \delta^b_a. \quad (21.21)$$

Then

$$\varepsilon^{\mu\nu\rho\sigma}\varepsilon_{abcd} = e e^\mu_{[a} e^\nu_b e^\rho_c e^\rho_{d]}, \quad e \equiv \det(e^a_\mu), \quad (21.22)$$

so that equations (21.21), (21.22) inserted into equation (21.18), yield, after a rescaling of the action

$$S_{\text{HE}} = \frac{1}{1024\pi G\lambda^2} S(Q^{(\varepsilon)}) = -\frac{1}{16\pi G}\int \mathrm{d}^4x e(R + 2\Lambda). \quad (21.23a)$$

Here

$$\left.\begin{aligned} R &= R^{ab}_{\mu\nu} e^\mu_a e^\nu_b, \\ \Lambda &= -12\lambda^2, \end{aligned}\right\} \quad (21.23b)$$

G = Newton's gravitational constant, and λ imaginary (real) corresponds to the de-Sitter (anti-de-Sitter) case.

From equations (21.20), (21.23) we recognize S_{HE} as the familiar Hilbert–Einstein action with cosmological term in *vierbein* notation (with the metric $g_{\mu\nu} = e^a_\mu e^b_\nu \eta_{ab}$, where η_{ab} is the Minkowski tensor, one can also write $e = \sqrt{(-g)}$ where $g = \det(g_{\mu\nu})$). The de-Sitter → Poincaré contraction described in chapter 4 corresponds to $\lambda \to 0$, so that in the Poincaré limit the scalar curvature density term is all that survives, the cosmological term vanishes. The field equations are most readily come by, using equation (21.8). Variation of the vierbein leads to the Einstein equations with cosmological term. Variation of the spin-connection $\omega^{ab}_\mu(\equiv h^{[ab]}_\mu)$ in turn produces the field equations

$$\bar{R}^a_{\mu\nu} = 0. \quad (21.24)$$

Using equations (21.2) and the structure constant expressions

$$f^a_{bc} = f^a_{[bc][de]} = 0, \quad f^a_{b[cd]} = -f^a_{[cd]b} = \tfrac{1}{2}(\eta_{bc}\delta^a_d - \eta_{bd}\delta^a_c), \quad (21.25)$$

we then recast equation (21.13) (in the e^a_μ, ω^{ab}_μ notation) in the form

$$\bar{R}^a_{\mu\nu} \equiv T^a_{\mu\nu} \equiv D_\mu e^a_\nu - D_\nu e^a_\mu = 0, \quad D_\mu e^a_\nu \equiv \partial_\mu e^a_\nu - \omega^{ab}_\mu e^c_\nu \eta_{bc} \quad (21.26)$$

These equations, algebraic in ω_μ^{ab} can be solved for ω_μ^{ab} and yield the torsionless Christoffel connection

$$\left.\begin{aligned} \omega_\mu^{ab} &= -(c_\mu^{ab} - c_{b\mu}^a - c_{\mu a}^b) \\ c_{\mu\nu}^a &= \partial_\mu e_\nu^a - \partial_\nu e_\mu^a,\ c_\mu^{ab} = e_\mu^f e_d^\rho e_e^\sigma \eta_{fc} \eta^{da} \eta^{eb} c_{\rho\sigma}^c, \text{etc.}\ldots \end{aligned}\right\} \tag{21.27}$$

as mentioned at the beginning of this section. We can use equations (21.27) to go to the second order formalism involving only the e_μ^a fields. Consider this substitution performed, and let us see now the effect on the action of an $sp(4)$ transformation $\varepsilon^a(x)$ which is not a Lorentz transformation. From equation (21.7), using the form (21.14) of Q_{AB} and the structure constants (21.25) we find the variation of the action to be proportional to $\bar{R}_{\mu\nu}^a$s (i.e., the torsion components) which now vanish. So at this level full $sp(4)$ invariance has been checked.

This construction of the Hilbert–Einstein action highlights one aspect of gravity theory usually left unmentioned. Were it not for the assumed space-inversion invariance (assumption (ii)) we need not have set $\beta = 0$ in the ansatz (21.12) and we could have had a parity violating gravity. This would have entailed solutions of the gravitational field equations without definite space-inversion properties. On the other hand, such solutions are possible even for $\beta = 0$. For instance the most general Lorentz invariant ansatz for e_μ^a and ω_μ^{ab} is

$$\left.\begin{aligned} e_\mu^a &= c(x^2)\delta_\mu^a + d(x^2)x_\mu x^a \\ \omega_\mu^{ab} &= f(x^2)(x^a\delta_\mu^b - x^b\delta_\mu^a) + g(x^2)\varepsilon^{ab}{}_{\mu\lambda}x^\lambda \end{aligned}\right\} \tag{21.28a}$$

with $x^2 \equiv (x^1)^2 + (x^2)^2 + (x^3)^2 - (x^0)^2$, $x_\mu \equiv \eta_{\mu\nu}x^\nu$. Even with $\beta = 0$ we can find solutions with both f and g nonvanishing. The simplest such solution

$$c(x^2) = C/(x^2)^{1/2}, d(x^2) = -C/x^2\sqrt{(x^2)}, f(x^2) = F/x^2, g(x^2) = G/x^2, \tag{21.28b}$$

with C, F, G constants, has the added features of a singularity at $x^2 = 0$ and of a metric

$$g_{\mu\nu} = e_\mu^a e_\nu^b \eta_{ab} = \frac{C^2}{x^2}\left(\eta_{\mu\nu} - \frac{x_\mu x_\nu}{x^2}\right) \tag{21.28c}$$

($x_\mu \equiv \eta_{\mu\nu}x^\nu$ as above), which is not invertible: $g^{\nu\rho}$ does not exist. Unlike in Einstein's theory, starting from the action (21.15) does not require the Riemannian invertibility of the metric. In all these senses (21.15) is wider in scope than the ordinary Hilbert–Einstein formulation. Incidentally the solution (21.28) has torsion

$$T^a_{\mu\nu} = -\frac{2CG}{x^2}\varepsilon^a{}_{\mu\nu\rho}\frac{x^\rho}{(x^2)^{1/2}} \tag{21.28d}$$

produced by an interference between parity violating and parity conserving amplitudes. Parity violation and torsion go hand-in-hand. Independently of any more realistic parity violating solution of the gravity equations this raises the cosmological question whether the universe as a whole is in a space-inversion symmetric configuration.

We have discussed these matters chiefly in order to stress the differences between the Hilbert–Einstein action and the action (21.15). In the next chapter we have an easy job repeating in a supersymmetric context the steps that led us to the action (21.15).

22
$N = 1$ Supergravity

To extend the construction given in the last section to the supersymmetric case (MacDowell & Mansouri 1977, Chamseddine & West 1977), we start from the anti-de-Sitter super-algebra $\mathscr{osp}(1|4)$ instead of the anti-de-Sitter algebra $\mathscr{sp}(4)$ ($\equiv \mathscr{osp}(0|4)$). The index A can now take four Majorana–Fermi values $\alpha = 1, 2, 3, 4$ in addition to the ten Bose values (21.11):

$$A \equiv (a = 0, 1, 2, 3) \oplus ([aa'] = [0\,1], [0\,2], [0\,3], [1\,2], [1\,3], [2\,3]) \oplus (\alpha = 1, 2, 3, 4). \tag{22.1}$$

We again define the structure constants, through the graded brackets of the superalgebra generators

$$[G_A, G_B] = f^C_{AB} G_C. \tag{22.2}$$

Subject to the Lorentz and space-inversion invariance constraints (i) and (ii) of the last section we now find

$$Q^{\mathrm{SUSY}}_{AB}(\chi) = \begin{cases} \varepsilon_{aa'bb'} & \text{for } A = [aa'], B = [bb'] \\ \chi(C\gamma_5)_{\alpha\beta} & \text{for } A = \alpha,\ B = \beta \\ 0 & \text{otherwise} \end{cases} \tag{22.3}$$

where, as in chapter 20, we set the parameter α multiplying $\varepsilon_{aa'bb'}$ equal to one, and χ is a numerical parameter. The $\mathscr{osp}(1|4)$ curvatures $\bar{R}^A_{\mu\nu}$ are now related to the super-Poincaré curvatures by

$$\left.\begin{aligned} \bar{R}^{ab}_{\mu\nu} &= R^{ab}_{\mu\nu} + h^c_\mu h^d_\nu f^{[ab]}_{cd} + h^\alpha_\mu h^\beta_\nu f^{[ab]}_{\alpha\beta} \\ \bar{R}^{\alpha}_{\mu\nu} &= R^{\alpha}_{\mu\nu} + (h^a_\mu h^\beta_\nu - h^a_\nu h^\beta_\mu) f^{\alpha}_{a\beta} \\ \bar{R}^{a}_{\mu\nu} &= \bar{R}^{a}_{\mu\nu} + h^\alpha_\mu h^\beta_\nu f^{a}_{\alpha\beta}. \end{aligned}\right\} \tag{22.4a}$$

Here $R^{ab}_{\mu\nu}$ are the ordinary (Lorentz) curvatures given by equation (21.20), $\bar{R}^a_{\mu\nu}$ the torsions of equation (21.26) and the new Fermi curvatures are

$$R^{\alpha}_{\mu\nu} = \partial_\mu h^\alpha_\nu - \partial_\nu h^\alpha_\mu + f^{\alpha}_{[ab]\gamma}(h^{[ab]}_\mu h^\gamma_\nu - h^{[ab]}_\nu h^\gamma_\mu). \tag{22.4b}$$

The ansatz (22.3) then yields

$$S(Q^{\mathrm{SUSY}}(\chi)) = \int \mathrm{d}^4x\, \varepsilon^{\mu\nu\rho\sigma}(\bar{R}^{ab}_{\mu\nu}\bar{R}^{cd}_{\rho\sigma}\varepsilon_{abcd} + \chi \bar{R}^{\alpha}_{\mu\nu}\bar{R}^{\beta}_{\rho\sigma}(C\gamma_5)_{\alpha\beta}). \tag{22.5}$$

Inserting here the expressions (22.4), we find

$$S(Q^{\text{SUSY}}(\chi)) = \int d^4x(T + K + C + E), \tag{22.6a}$$

with

$$T = \varepsilon^{\mu\nu\rho\sigma}[\varepsilon_{abcd}(R^{ab}_{\mu\nu}R^{cd}_{\rho\sigma} + 2R^{ab}_{\mu\nu}h^{\alpha}_{\rho}h^{\beta}_{\sigma}f^{[cd]}_{\alpha\beta}) + \chi R^{\alpha}_{\mu\nu}R^{\beta}_{\rho\sigma}(C\gamma_5)_{\alpha\beta}] \tag{22.6b}$$

$$K = \varepsilon^{\mu\nu\rho\sigma}(2\varepsilon_{abcd}R^{ab}_{\mu\nu}h^{e}_{\rho}h^{f}_{\sigma}f^{[cd]}_{ef} + 4\chi R^{\alpha}_{\mu\nu}h^{a}_{\rho}h^{\gamma}_{\sigma}f^{\beta}_{a\gamma}(C\gamma_5)_{\alpha\beta}) \tag{22.6c}$$

$$C = \varepsilon^{\mu\nu\rho\sigma}(\varepsilon_{abcd}h^{a'}_{\mu}h^{b'}_{\nu}h^{c'}_{\rho}h^{d'}_{\sigma}f^{[ab]}_{a'b'}f^{[cd]}_{c'd'} + 2\varepsilon_{abcd}h^{a'}_{\mu}h^{b'}_{\nu}h^{\alpha}_{\rho}h^{\beta}_{\sigma}f^{[ab]}_{a'b'}f^{[cd]}_{\alpha\beta}$$
$$+ 4\chi h^{a}_{\mu}h^{\gamma}_{\nu}h^{b}_{\rho}h^{\delta}_{\sigma}f^{\alpha}_{a\gamma}f^{\beta}_{b\delta}(C\gamma_5)_{\alpha\beta}) \tag{22.6d}$$

$$E = \varepsilon^{\mu\nu\rho\sigma}\varepsilon_{abcd}h^{\alpha}_{\mu}h^{\beta}_{\nu}h^{\gamma}_{\rho}h^{\delta}_{\sigma}f^{[ab]}_{\alpha\beta}f^{[cd]}_{\gamma\delta}. \tag{22.6e}$$

Besides the old structure constants (21.17), (21.19), (21.25) of the anti-de-Sitter algebra, with λ real, the extra structure constants of the corresponding superalgebra appear here. In a basis, chosen to agree with the notations of Deser & Zumino (1976), they are

$$\left.\begin{aligned} f^{[ab]}_{\alpha\beta} &= 2i\lambda(C\sigma^{ab})_{\alpha\beta} \\ f^{a}_{\alpha\beta} &= \tfrac{1}{2}i(C\gamma^{a})_{\alpha\beta} \\ f^{\gamma}_{a\beta} &= \lambda(\gamma_a)^{\gamma}{}_{\beta} \\ f^{\gamma}_{[ab]\beta} &= -\tfrac{1}{2}(\sigma_{ab})^{\gamma}{}_{\beta}. \end{aligned}\right\} \tag{22.7a}$$

Making use of the identity

$$\varepsilon_{abcd}f^{[cd]}_{\alpha\beta} = 8i\lambda f^{\gamma}_{[ab]\alpha}(C\gamma_5)_{\gamma\beta}, \tag{22.7b}$$

one readily checks that the last two terms in the expression for T (equation (22.6*b*)) assemble into an exact divergence for the special value

$$\chi = 8i\lambda \tag{22.8}$$

of the parameter χ (the terms quadratic in derivatives of the Fermi fields are, by themselves, already an exact divergence; the terms linear in derivatives of the Fermi fields from the expansion of the last term in T, along with those linear in derivatives of the spin connection $h^{[ab]}_{\mu}$ from the expansion of the second term in T, add up to an exact divergence once (22.8) is imposed; the nonderivative terms originating from the last two terms in T cancel). In particular this remarkable set of cancellations allows the Fermi fields to obey first order field equations. The first term in T is an exact divergence (the Euler number density) as already noticed in chapter 21. Thus, all of T is an exact divergence, which contributes a topological surface term to the action. Without any further ado then, T can be ignored.

The terms K and C are the supersymmetrizations of the Hilbert–Einstein lagrangian and of the cosmological term respectively, whereas the term E vanishes identically.

With the conventional notations (G = Newton's gravitational constant)

$$h^a_\mu \equiv e^a_\mu,\ h^{[ab]}_\mu \equiv \omega^{ab}_\mu,\ h^\alpha_\mu \equiv (8\pi G)^{1/2}\psi^\alpha_\mu \tag{22.9}$$

the $OSp(1|4)$ supergravity action can be written, up to surface terms, in the form (Deser & Zumino 1976)

$$\begin{aligned} S_{\mathrm{SG}}(\lambda) &= \frac{1}{1024\pi G\lambda^2} S(Q^{\mathrm{SUSY}}(\chi = 8\mathrm{i}\lambda)) \\ &= S_{\mathrm{SG}}(\lambda = 0) - S_{\mathrm{SUSY\ cosmological}} \end{aligned} \tag{22.10a}$$

where

$$\left.\begin{aligned} S_{\mathrm{SG}} \equiv S_{\mathrm{SG}}(\lambda = 0) &= \int \mathrm{d}^4x\left(-\frac{1}{16\pi G}eR - \tfrac{1}{2}\mathrm{i}e^{\mu\nu\rho\sigma}\bar{\psi}_\mu\gamma_5\gamma_\nu D_\rho\psi_\sigma\right) \\ D_\mu = \partial_\mu - \tfrac{1}{2}\omega^{ab}_\mu\sigma_{ab} &\quad \sigma_{ab} = \tfrac{1}{4}[\gamma_a, \gamma_b]- \\ [D_\mu, D_\nu] &= -\tfrac{1}{2}R^{ab}_{\mu\nu}\sigma_{ab} \\ R = R^{ab}_{\mu\nu}e^\mu_a e^\nu_b, &\quad e = \det(e^a_\mu), \end{aligned}\right\} \tag{22.10b}$$

the $R^{ab}_{\mu\nu}$ are given as before by the equation (21.20),

$$S_{\mathrm{SUSY\ cosmological}} = \int \mathrm{d}^4x\left(-\frac{1}{8\pi G}\Lambda e - \tfrac{1}{2}\mathrm{i}m\varepsilon^{\mu\nu\rho\sigma}\bar{\psi}_\mu\gamma_5\sigma_{\nu\rho}\psi_\sigma\right) \tag{22.10c}$$

and as before

$$\left.\begin{aligned} \Lambda &= -12\lambda^2, \\ m &= 2\lambda = (-\Lambda/3)^{1/2} \end{aligned}\right\} \tag{22.10d}$$

with λ real, corresponding to the anti-de-Sitter algebra $\mathfrak{sp}(4)$.

Upon anti-de-Sitter → Poincaré contraction ($\lambda \to 0$), the supersymmetrized cosmological term, equation (22.10c) (Das & Freedman 1977, Townsend 1977, MacDowell & Mansouri 1977) drops out and one recovers the action (22.10b) of Poincaré supergravity, the supersymmetrized Einstein action, in close similarity to what happened in the nonsupersymmetric case of chapter 21.

We now proceed to a more detailed discussion of the $\lambda \to 0$ Poincaré supergravity and will return to the cosmological term at the end of this chapter. For the discussion at hand, we consider the action S_{SG} of equation (22.10b). It is convenient at this point to adopt units such that $8\pi G = 1$.

The field equations corresponding to variations of the vierbein e^a_μ are then

$$G^{\lambda\nu} = \tfrac{1}{2}\mathrm{i}\varepsilon^{\mu\nu\rho\sigma}\bar{\psi}_\mu\gamma_5\gamma^\lambda D_\rho\psi_\sigma \tag{22.11a}$$

where

$$G^{\lambda\nu} = e^\nu_a(R^\lambda_a - \tfrac{1}{2}e^\lambda_a R) \tag{22.11b}$$

and

$$R_{\nu a} = R_{\nu\lambda a}{}^{\lambda} \tag{22.11c}$$

is the (nonsymmetric) Ricci tensor. Recalling the expression for the torsion $\bar{R}^a_{\mu\nu}$ (equation (21.26))

$$\bar{R}^a_{\mu\nu} \equiv T^a_{\mu\nu} = D_\mu e^a_\nu - D_\nu e^a_\mu,\ D_\mu e^a_\nu = \partial_\mu e^a_\nu - \omega_\mu{}^a{}_b e^b_\nu, \tag{22.11d}$$

the field equations obtained by varying the spin-connection ω^{ab}_μ take the form

$$T^\rho_{\mu\nu} \equiv T^a_{\mu\nu}e^\rho_a = \tfrac{1}{2}\mathrm{i}\bar{\psi}_\mu\gamma^\rho\psi_\nu \tag{22.11e}$$

and finally the field equations for the spin–vector Rarita–Schwinger gravitino field ψ^a_μ are

$$eE^\mu \equiv \varepsilon^{\mu\nu\rho\sigma}(\gamma_\nu D_\rho - \tfrac{1}{4}\gamma_\lambda T^\lambda_{\nu\rho})\psi_\sigma = 0. \tag{22.11f}$$

Throughout these equations

$$\gamma_\nu = e^a_\nu\gamma_a \quad \gamma^\nu = e^\nu_a\gamma_b\eta^{ab} \tag{22.11g}$$

with γ_a the ordinary Dirac matrices. The supersymmetry transformations

$$\left.\begin{aligned} \delta e^a_\mu &= \mathrm{i}\bar{\varepsilon}\gamma^a\psi_\mu \\ \delta\psi_\mu &= 2D_\mu\varepsilon \\ \delta\omega^{ab}_\mu &= B^{ab}_\mu - \tfrac{1}{2}e^b_\mu B^{ac}_c + \tfrac{1}{2}e^a_\mu B^{bc}_c \end{aligned}\right\} \tag{22.12a}$$

where

$$\left.\begin{aligned} B^{\lambda\mu}_a &= \mathrm{i}\bar{\varepsilon}\gamma_5\gamma_a D_\nu\psi_\rho\varepsilon^{\lambda\mu\nu\rho} \\ B^{ab}_\mu &= B^{\rho\sigma}_c e^a_\rho e^b_\sigma e^c_\mu \quad B^{ab}_c = e^\mu_c B^{ab}_\mu \end{aligned}\right\} \tag{22.12b}$$

and $\varepsilon(x)$ is the local Majorana-spinor 'Fermi parameter', leave the action (22.10*b*) invariant. The transformations (22.12) close *on* the mass-shell but not off it; auxiliary fields have not yet been included.

The novel and surprising feature of all this is that we encounter a natural interacting Rarita–Schwinger field. But as a rule such interacting Rarita–Schwinger fields cause all kinds of problems. What happens is that by taking the covariant divergence of the Rarita–Schwinger equation (22.11*f*) one obtains in general a new independent constraint,

$$D_\mu(eE^\mu) = 0 \tag{22.13}$$

which when imposed on the theory causes inconsistencies and acausal propagation (Velo & Zwanziger 1969).

In supergravity, equation (22.13) is *not* an independent constraint, rather it follows from the field equations and local supersymmetry, as we now show. Consider for this purpose a general variation of the action (22.10*b*)

$$\delta S_{\text{SG}} = \int d^4x \left(\frac{\delta S_{\text{SG}}}{\delta e^a_\mu} \delta e^a_\mu + \frac{\delta S_{\text{SG}}}{\delta \omega^{ab}_\mu} \delta \omega^{ab}_\mu + \frac{\delta S_{\text{SG}}}{\delta \psi_\mu} \delta \psi_\mu \right). \tag{22.14}$$

If, now, the variations of the fields are of the local supersymmetry type (22.12), then the last term in (22.14) becomes

$$2 \int d^4x (D_\mu \bar{\varepsilon}) e E^\mu \tag{22.15}$$

where we used the fact that E^μ is precisely $\delta S_{\text{SG}}/\delta \psi_\mu$. A partial integration then recasts (22.15) in the form

$$-2 \int d^4x \bar{\varepsilon} D_\mu (e E^\mu). \tag{22.16}$$

If we now impose the e^a_μ and ω^{ab}_μ field equations then the first two terms in (22.14) are automatically zero, so that with (22.16), (22.14) becomes

$$\int d^4x \bar{\varepsilon} D_\mu (e E^\mu) = 0. \tag{22.17}$$

The parameter $\varepsilon(x)$ being arbitrary, this means

$$D_\mu (e E^\mu) = 0 \tag{22.18}$$

which is precisely the potentially troublesome constraint (22.13). So, this constraint follows from local supersymmetry and the bosonic field equations, and as such can cause no new problems. Had we placed the Rarita–Schwinger field in a massless $(\frac{3}{2}, 1)$ supermultiplet involving one spin $\frac{3}{2}$ and one spin 1 field, (the existence of such a supermultiplet was proved in chapter 5), no *local* supersymmetry would have been available (without spin 2 the Poincaré part remains ungauged) and the constraint (22.13) would not have been defused. It is for this reason that such $(\frac{3}{2}, 1)$ supermultiplets are never taken into consideration. Even letting them supergravitate would only defuse the type (22.17) gravitino constraints, not those of these additional spin $\frac{3}{2}$ fields in $(\frac{3}{2}, 1)$ supermultiplets, unless further $N > 1$ local supersymmetries were to be enforced.

At this point it is worthwhile considering in some more detail the supersymmetrized cosmological term of equation (22.10*c*) which we contracted away ($\lambda \to 0$). The first term in $S_{\text{SUSY cosmological}}$ is of course just the ordinary cosmological term, whereas the second term looks like a gravitino mass term. This seems paradoxical since the gravitino is in the same supermultiplet with the *massless* graviton. The paradox is lifted (Deser & Zumino 1977) when we realize that with a nonvanishing cosmological constant we have to quantize in an anti-de-Sitter rather than a Minkowski background, and that in such a background 'masslessness' is a subtle concept. It can be defined for nonvanishing spin, by requiring an appropriate gauge invariance that reduces the number of degrees of freedom to two. As a matter of fact, the local supersymmetry transformations which leave invariant the full action (22.10*a*) with $\lambda \neq 0$, differ from those given in equation (22.12*a*) in that the gravitino field transforms as

$$\delta\psi_\mu^\alpha = 2(D_\mu + \tfrac{1}{2}m\gamma_\mu)\varepsilon. \tag{22.19}$$

This insures the proper gravitino 'masslessness' in an anti-de-Sitter background.

Having abandoned superspace two chapters ago, we have also forfeited our rights to a clear and immediate picture of off-shell local supersymmetry with the *de rigueur* auxiliary fields. Off-shell, all 16 components of the Rarita–Schwinger field must be matched. The vierbein can always have 6 of its 16 components removed by a local Lorentz transformation, leaving an imbalance of at least 6 Bose fields. This can be remedied in a 'minimal' way by introducing an axial vector field (four components) and two scalars (two components) (Ferrara & van Nieuwenhuizen 1978, Stelle & West 1978). However this is not unique and there exist further possible sets of auxiliary fields (Breitenlohner 1977, Siegel 1979, Sohnius & West 1981). The coupling of this supergravity multiplet to matter and gauge supermultiplets (Cremmer *et al.*, 1979, Cremmer, Ferrara, Girardello & van Proeyen 1983, Bagger & Witten 1983) is of importance in the phenomenology of supersymmetric grand unification (Overt & Wess 1982, Nilles 1984).

23
Extended supergravities

From table 5.1 it can be seen that in the case of extended supersymmetry of type N, a supermultiplet *must* contain states of helicity h such that $|h| \geqslant N/4$. This means that for $N > 2$ we have no supermultiplets with scalars and spin one-half fermions only; for $N > 4$ we must exceed helicity one, (thus no Yang–Mills supermultiplets for $N > 4$) and as of $N \geqslant 9$ states of helicity larger than two make an appearance (the critical values $N = 2$, 4, 8 correspond to the CPT self-conjugate cases mentioned in chapter 5). There are strong reasons to believe (though really no proof as yet) that nontrivial interacting theories containing fields of spin five-halves and higher in the lagrangian are inconsistent (Aragone & Deser 1979, Curtright 1979 Berends, van Holten, de Wit & van Nieuwenhuizen 1980). Hence one is limited to supergravities with $N \leqslant 8$.

The 'natural' $N = 1$ supergravity in four space–time dimensions has just been presented. Here we ask for the extended supergravities ($N > 1$). The corresponding lagrangians can be constructed. We shall not do so here, but will content ourselves with a few remarks on these lagrangians and on the phenomenology of the $N = 8$ theory. First of all, the elegant construction (MacDowell & Mansouri 1977) we presented for the $N = 1$ case, already fails at the $N = 2$ level, as in addition to the types of terms in the ansatz (22.5) it requires terms explicitly containing the vierbein or metric (Townsend & van Nieuwenhuizen 1977). Fortunately the additional type of term is simple and is readily guessed.

An alternative method starts from the Noether current of global supersymmetry, couples this to the gravitino and then introduces extra generally covariant terms with arbitrary coefficients which are then determined so that local supersymmetry is obtained. In a sense this is the supersymmetric generalization of the methods of Gupta (1954), Thirring (1961), and Feynman (1963), from ordinary gravity theory.

Another technique first constructs natural ($N = 1$) supergravities in space–times of dimensionality $d > 4$ and then dimensionally reduces them to $d = 4$. If the small manifold in the extra dimensions is sufficiently symmetrical, extended supergravities in $d = 4$ are obtained. We shall see more on this in the next two chapters. The idea here is that constructing natural (i.e., $N = 1$) supergravities is always simpler. Yet even at this level it must be admitted that general methods do not exist. Fortunately there exists a maximal supergravity in $d = 11$ dimensions from which all four-dimensional supergravities can be retrieved.

Superspace methods would seem then to be indicated in order to eliminate all this guesswork. Unfortunately, even for low N these methods are way too cumbersome to recommend themselves with any degree of finality. For $N > 2$ supergravities superspace methods are not available and reasonable doubts have been entertained concerning the existence of auxiliary fields for these theories.

We now turn to a brief discussion of the phenomenological possibilities of extended supergravities. In extended supergravity of type N the largest internal nonabelian gauge group is $O(N)$, corresponding to a gauged $OSp(N|4)$. Were we to partially or fully contract, some or all of this gauge group would be lost in favor of gauged central charges (see chapter 4). The phenomenologically requisite $SU(3)_{\text{color}}$ gauge group will thus be available inside $O(N)$ only for $N \geqslant 6$. Requiring at least one additional $U(1)$ symmetry, say for electromagnetism, further pushes N up to 7 or 8. Since the $N = 7$ and $N = 8$ theories are indistinguishable by the time the $N = 7$ theory's spectrum is CPT doubled, we will concentrate exclusively on the maximal CPT self-conjugate $N = 8$ theory. There are a number of variants of this theory. The largest nonabelian gauge symmetry is $O(8)$ (de Wit & Nicolai 1982) but there are versions with a product of 28 abelian gauge groups (Cremmer & Julia 1979) or a noncompact nonabelian gauge group (Hull 1984). We shall consider here the de Wit–Nicolai version with gauged $O(8)$. An interesting phenomenon (already noted in the Cremmer–Julia version) is the appearance of a 'hidden' gauged $SU(8)$ symmetry. One may wonder how this is possible, there being only 28 massless vector bosons in the $N = 8$ supergravity multiplet described in table 5.1 whereas a local $SU(8)$ symmetry would require 63 massless vector particles. The answer is that the gauging of $SU(8)$ is done nonlinearly through 'composite' combinations of scalar fields. To understand how this can be, consider the following example (Witten 1979). Let $\phi^1, \dots . \phi^N$ be N complex scalar fields coupled to the abelian gauge field A_μ in a gauge invariant way as in the lagrangian

$$\mathscr{L}_\lambda = (\partial_\mu + ieA_\mu)\phi^{\dagger a}(\partial_\mu - ieA_\mu)\phi^a - \tfrac{1}{4}\lambda F_{\mu\nu}F^{\mu\nu} \tag{23.1}$$

where summation over a from 1 to N is implied, $F_{\mu\nu} = \partial_\mu A_\nu - \partial_\nu A_\mu$ and λ is a parameter usually set equal to 1 in Maxwell theory. The Maxwell equation is

$$-\lambda\partial^{\mu}F_{\mu\nu}=ie\phi^{\dagger a}\overleftrightarrow{\partial}^{\nu}\phi^{a}+2e^{2}A_{\mu}\phi^{\dagger a}\phi^{a}. \tag{23.2}$$

Now let the parameter λ go to zero. The A_ν field becomes nonpropagating and (23.2) becomes a set of four algebraic equations for A_ν with the solution

$$A_{\nu}=\frac{-i\phi^{\dagger a}\overleftrightarrow{\partial}^{\nu}\phi^{a}}{2e\phi^{\dagger b}\phi^{b}}. \tag{23.3}$$

Inserting (23.3) into $\mathscr{L}_{\lambda=0}$ we find a new, still gauge invariant, lagrangian that does not contain the vector field A_ν anymore, but which is highly nonlinear. The gauging is done by the scalar field composite (23.3). This argument is readily generalized to the nonabelian case, and this is the way $SU(8)$ is gauged in $N=8$ supergravity. In the fully contracted Cremmer–Julia case the 70 scalars of the supergravity multiplet span the coset space $E_{7,7}/SU(8)$. Here $E_{7,7}$ is the maximally noncompact real form of the Lie group E_7, the one whose maximal compact subgroup is $SU(8)$. The group $E_{7,7}$ has dimension 133, $SU(8)$ has dimension 63 so that the coset space has 70 dimensions as required. In this case $E_{7,7}$ is a global symmetry of the theory. In the de Wit–Nicolai case, in which $O(8)$ is gauged, the local $SU(8)$ is again present, though not the global $E_{7,7}$ symmetry. The theory thus has local $SO(8)\times SU(8)$ symmetry to start with, which is then broken to $SO(8)$ by $SU(8)$ gauge fixing. The low energy gauge group $SU(3)_{\text{color}}\times(SU(2)\times U(1))_{\text{electroweak}}$ fits inside $SU(8)\times SO(8)$, though certainly not inside $SO(8)$. Anyway, the $SO(8)$ can further be broken according to the shape of the scalar potential. This potential is fully determined in the theory and its critical points have been studied (Warner 1984). It has a stable critical point that preserves the full $SO(8)$ symmetry, a stable critical point where $SO(8)$ breaks down to $SU(3)\times U(1)$, one where it breaks down to G_2, two unstable critical points with $SO(7)$ residual symmetry, one critical point with $SO(6)$ left over (stability unknown), as well as critical points for which the leftover symmetry cannot fit even $SU(3)$ (e.g., $SO(3)\times SO(3)$).

In one word, the symmetry breaking situation is complicated, as is the $N=8$ theory itself. As we shall see it is much more profitable to start from a simple higher-dimensional theory, which puts all this complexity into a neat perspective (chapter 25). Were we to take phenomenologically seriously the gauged $SO(8)$, we would have to concentrate on its maximal $SU(3)\times U(1)\times U(1)$ subgroup and attempt to identify the $SU(3)$ factor with the color gauge group, one of the $U(1)$ factors with the Maxwell–Weyl gauge group of electromagnetism, and the other $U(1)$ either with the 'third' component of Glashow's weak $SU(2)$ or with some yet unknown abelian gauge group. In either case the charged weak gauge bosons $W^{\pm}$, at the very least, would not fit into $SO(8)$. They would have to be composite, qualitatively different from the gluon, the photon, and possibly even the Z^0. This is disappointing.

Let us now see what can be said about the fermions (Gell-Mann 1977). Concentrating for the time being on the supergravity multiplet from table 5.1, we notice the presence of an $SO(8)$ $\mathbf{8}$ of massless spin three-halves gravitini and of a $\mathbf{56}$ (Young tableau $\begin{array}{|c|}\hline \ \\ \hline \ \\ \hline \ \\ \hline\end{array}$) of massless spin one-half fermions. The interesting branching of the $SO(8)$ octet with respect to $SU(3) \times U(1) \times U(1)$ is $\mathbf{8} = \mathbf{3} + \bar{\mathbf{3}} + \mathbf{1} + \mathbf{1}$ where, for the moment, we suppress the $U(1) \times U(1)$ attributes and only indicate the $SU(3)$ multiplicities. We then readily find the branching rule for the $\mathbf{56} = \mathbf{1} + \mathbf{1} + \mathbf{1} + \mathbf{1} + \mathbf{3} + \bar{\mathbf{3}} + \mathbf{3} + \bar{\mathbf{3}} + \mathbf{3} + \bar{\mathbf{3}} + \mathbf{3} + \bar{\mathbf{3}} + \mathbf{6} + \bar{\mathbf{6}} + \mathbf{8} + \mathbf{8}$. Of these 56 'spin' one-half states, eight $(\mathbf{3} + \bar{\mathbf{3}} + \mathbf{1} + \mathbf{1})$ get absorbed as the helicity $\pm$ one-half states of *massive* gravitini, via a fermionic Higgs phenomenon. Alongside the quark–antiquark-like triplets, and 'exotic' (sextet and octet) quarks, this leaves but *two* color singlet leptons, thus frustrating any phenomenology. The electric charges can be arranged so that the color triplets have the usual fractional charges and the leptons have integral charges, but as we just saw, that is small consolation.

If one wants to achieve any amount of realism one has to escape the constraints of table 5.1. To this end one can reinterpret (Curtright & Freund 1979) all or part of the supergravity multiplet as preons (Pati & Salam 1974), some kind of more fundamental building blocks than the quarks and leptons. Just like for the $W^{\pm}$, we could then search also for the Z^0, the gluons (?), the Higgs-bosons, quarks and leptons among composites (bound states) of $N = 8$ supergravity. The dynamical problem of finding the bound state spectrum is formidable indeed. Therefore, all kinds of phenomenological approaches were developed. Cremmer & Julia (1979) proposed that the originally nonlinearly gauged $SU(8)$ develops propagating vector gauge bosons in quantum theory. A similar phenomenon is known to occur in two-dimensional σ-models (Witten 1979). Centered on this idea's combinations with supersymmetry, an $SU(8)$ phenomenology was born (Ellis, Gaillard & Zumino 1980) which we do not present here in detail. In the context of higher-dimensional theories similar problems arise, and one is led to reconsider this phenomenology.

It is gratifying that a tinge of realism – inclusion of color, fractional quark charges, and integral lepton charges – can be perceived at even the most naive level. In addition, $SU(8) \supset SU(5)_{\text{grand unification}} \times SU(3)_{\text{generation}} \times U(1)$ leading naturally (Curtright & Freund 1979) to an $SU(3)_{\text{generation}}$ which suggests three generations, just as seen until now in experiment. But all this makes the ultimate absence of a compelling and realistic spectrum all the more frustrating. There is no question that more work along these lines, but with a higher-dimensional starting point, is forthcoming.

24

The hidden assumptions of grand unification and the matter/force problem

There are some features of $N = 8$ supergravity, shared, as we shall see, with higher-dimensional supergravities, that make these theories so very attractive. First of all, there are some assumptions in both ordinary and supersymmetric grand unification, so well hidden, that it is usually glossed over that they even are assumptions. Any grand unification starts by naming the grand unifying compact simple gauge group $G(SU(5), SO(10)\ldots)$. Now a clever guess concerning G goes a long way, but the list of simple compact Lie groups is infinite (Cartan 1894), and in theoretical physics a specific choice of G out of this infinite list ought to be *theoretically* justified. In ordinary grand unification no ideas for such a theoretical justification of the gauge group are ever alluded to. In extended ($N = 8$) supergravity we have seen that the choice of G is quite limited: $SO(8) \times SU(8)$ or some subgroup thereof as chosen by the extrema of the potential. Once G is chosen, the gauge bosons are uniquely specified, but grand unification again refers to inspired guesses when it comes to the assignment of the matter (spin one-half fermions + Higgs scalars) to specific G-multiplets. Again these assignments, modulo all the difficulties mentioned in chapter 23, are theoretically dictated in $N = 8$ supergravity.

The solution of a third problem lies then close at hand. Einstein (1955) repeatedly emphasized the conceptual imbalance between the two sides of his gravitational equations. On the left-hand side sits the Einstein tensor $G_{\mu\nu} = R_{\mu\nu} - \frac{1}{2}g_{\mu\nu}R$, a genuinely geometrical construct, whereas on the right-hand side we find the energy–momentum tensor totally unspecified by the theory. Its arbitrariness reflects our freedom to postulate all kinds of gravitating matter. Indeed this problem is a very old one, going back essentially to Newton. Two physical 'categories' are independently introduced: *matter* and *force*. But, *a priori*, there is no connection between the various forces, or between the various kinds of matter, let alone between the various kinds of forces and the various kinds of matter. In quantum field theory all forms of matter imply the existence of 'forces' corresponding to the exchange of the corresponding matter quanta. But even at this level, the exchange of fermions is qualitatively different from the exchange of bosons. Two 'objects' can interact via single boson exchange without losing their 'identities'. By contrast, this is not possible via single fermion exchange. Two fermion exchange forces do not require the loss of identity of the interacting objects, but such forces are qualitatively different from single boson exchange forces. The point of view that the fermions (quarks, leptons,...) are the basic forms of matter out of which more complex matter forms (baryons, mesons, nuclei...) are 'built' has therefore gained acceptance. Thus at the quantum field theory level the old Newtonian matter/force dichotomy survived. Forces, now related to Bose fields and their quanta (photons, gluons, weak bosons, gravitons...), came to be viewed, by and large, as providing the binding between the fermionic constituents of matter. Of course these bosonic quanta can themselves bind into rarer matter forms (glueballs, solitons,...). $N = 1$ supersymmetry in four space–time dimensions enforces a certain correlation between fermions and bosons, but this correlation is insufficient for solving the matter/force problem. Sure the graviton has its Fermi partner the gravitino, the gluon the gluino,..., the quark its Bose partner the squark,.... Yet, no correlation whatsoever is provided between the gauge force supermultiplets (graviton–gravitino, photon–photino,...) on the one hand, and the matter supermultiplets (quark–squark, lepton–slepton,...) on the other hand. The situation in $N = 8$ supergravity (in four space–time dimensions) is radically different. There exists but *one* supermultiplet that does not run over into helicities $\geqslant \frac{5}{2}$. This supermultiplet already contains the graviton. To the extent that we want a unique graviton, we are forced to start with *one* such $N = 8$ supergravity multiplet, to which *no* further forms of 'matter' or 'force' can be added. This $N = 8$ supergravity multiplet already contains gravity, chromodynamics and the electroweak(?) forces, all unified at the Planck scale, *as well* as a fully specified set of spin one-half ('matter') and spin zero ('Higgs') fields. All basic forces *and* all basic forms of matter now appear in the *same* supermultiplet. Newton's dichotomy is removed, Einstein's 'complaint' is answered. Just as the Einstein tensor $G_{\mu\nu}$ is specified by general covariance and local Lorentz invariance, so the form of the energy–momentum tensor is picked by local supersymmetry. The different basic forces together with the different basic forms of matter form a whole. They are all but different members of the same supermultiplet, different aspects of the same phenomenon. Just as a Lorentz transformation can switch electricity and magnetism, so a supersymmetry transformation can switch force and matter. No previous physical theory has exhibited anything like this degree of self-containedness and completeness. It is in view of all this that attempts at constructing a phenomenologically viable (remember all the difficulties of chapter 23) supergravity, or superstring theory are being intensely pursued at present. This brings us unambiguously to higher-dimensional theories.

25

Higher-dimensional unification

Consider two interesting four-dimensional supersymmetric theories which we have encountered before: the maximal $N=4$ Yang–Mills theory, and the maximal $N=8$ supergravity theory. Both these theories have remarkable properties: the $N=4$ Yang–Mills theory is finite, the $N=8$ supergravity solves the matter/force problem (it also may have improved convergence properties). Yet, as is clear from table 5.1 both these theories have rich particle spectra, and complicated lagrangians describe the manifold interactions of these numerous fields. For $N=8$ supergravity we even recorded the existence of various forms of the theory (Cremmer–Julia, de Wit–Nicolai, etc...). Somehow this all flies in the face of an unwritten rule of theoretical physics, namely that important theories be simple, unique and beautiful. Could it be that the apparent aesthetic flaws of these theories, are consequences of our way of looking at them, rather than of the theories themselves? What I have in mind is something like looking at an animal outside of its natural habitat when it can easily appear clumsy and weird. Only replacing it on its home ground will reveal its natural grace. What is the natural habitat of these theories?

Take the $N=4$ supersymmetric Yang–Mills theory in four space–time dimensions. Its particle spectrum involves all massless particles in the adjoint representation of the gauge group G; $\dim G$ spin one particles (each with two transverse degrees of freedom), $6\dim G$ scalars and $4\dim G$ spin one-half Majorana particles (two degrees of freedom each), a total of $8\dim G$ Bose degrees of freedom and $8\dim G$ Fermi degrees of freedom. Now imagine space–time had one time and nine space dimensions. A vector field would then have ten components, a massless vector field eight transverse degrees of freedom. The metric signature of this ten-dimensional space being eight, it admits Majorana–Weyl spinors also with eight degrees of freedom (see chapter 3). Next we construct the $N=1$ supersymmetric Yang–Mills theory (gauge group G) in ten-dimensional space–time. It amounts to the Yang–Mills term ($\sim F^2$) and to a minimal coupling of the G-gauge field to the adjoint G-representation Majorana–Weyl spinor, as simple a lagrangian as could be hoped for. If six of the space dimensions were to curl up into a six-torus T^6, Minkowski ten-space M_{10} would be replaced by the product $V_{10} = M_4 \times T^6$(M_4 = Minkowski four-space). All fields would be periodic in the six curled up dimensions. They would correspond to towers of four-dimensional particles of spin one, one-half and zero and ever increasing mass. The lowest mass in each tower would be zero (zeroth harmonics), and the mass scale for the higher excitations would be set by the radii of the six circles whose product is the six-torus. In a step usually referred to as *dimensional reduction* let all these six radii shrink down to zero. V_{10} would then lose six of its dimensions and become ordinary M_4. At the same time, all nonvanishing masses would go to infinity (the mass scales being inversely proportional to the circle radii), and thus be removed from the physical spectrum. One reaches a four-dimensional theory with only massless particles. A ten-dimensional vector field $A_M(M = 0, 1, 2, 3, 5, 6, \ldots, 10)$ appears in M_4 as a four dimensional vector field $A_\mu(\mu = 0, 1, 2, 3)$ and six scalar fields ($M = 5, \ldots, 10$), where we assumed the curled up dimensions to be $x^5, x^6, \ldots, x^{10}$. Similarly the eight degrees of freedom of the ten-dimensional spinor reorganize themselves into four four-dimensional Majorana spinors with two degrees of freedom each. The spectrum obtained this way is the same as that of the $N = 4$ supersymmetric Yang–Mills theory in four-dimensions. Following these reductions of the fields, the simple ten-dimensional lagrangian itself can be rewritten as a four-dimensional lagrangian and this turns out to be precisely that of the $N = 4$ supersymmetric Yang–Mills theory. The ten-dimensional Lorentz group $SO(9, 1)$ contains the product $SO(3, 1) \times SO(6)$ with $SO(3, 1)$ the Lorentz group of four-dimensional Minkowski space, and $SO(6)$ corresponding to rotations in the dimensions $x^5, \ldots, x^{10}$. The four-dimensional theory retains this $SO(6) \sim SU(4)$ invariance and this $SU(4)$ is what enters the $N = 4$ extended conformal supergroup $PSU(4|2, 2)$ of the $N = 4$ supersymmetric Yang–Mills theory in four dimensions. It is really remarkable that a quite involved four-dimensional theory is thus reproduced by dimensionally reducing a ten-dimensional theory of unusual simplicity. In all this discussion the ten-dimensional flat space served as a purely mathematical device, which simplified the four-dimensional theory. Could one however take the higher-dimensional space seriously and let the size of the – in this example – six-torus become small without actually going to zero? We are then confronted with the questions:

(I) what is the 'true' dimension of space–time?

(II) Is there a way of predicting the 'true' and the 'apparent' dimensions of space–time?

This brings us into the field of modern Kaluza–Klein theory (Appelquist, Chodos & Freund 1985).

Leaving these 'aesthetic' considerations aside, the most compelling reason to consider higher-dimensional theories comes from the need to construct a consistent quantum theory of gravity. Already in four space–time dimensions, gravity is nonrenormalizable. The only way to achieve quantum consistency for gravity then remains the possibility of a finite quantum theory: infinitely many counterterms may be needed but their coefficients are calculable and finite (small) rather than arbitrary constants. Even in the context of extended (all the way to $N = 8$) supergravity this does not seem to occur. This had led to an extension of the search to higher dimensions. Before we go into the details of the most interesting theories of this type, let us briefly review the general geometric framework of such a theory.

Basic formalism

I will present the formalism in the simplest give-dimensional case. The five-dimensional manifold M_5 is assumed of the form $M_5 = M_4 \times S^1$ with M_4 the 'ordinary' four-dimensional universe and S^1 a small circle. Neglecting higher harmonics (nonzero-modes) the metric is then of the form

$$\gamma_{MN} = \left(\begin{array}{c|c} g_{\mu\nu} + e^2\kappa^2\phi A_\mu A_\nu & e\kappa\phi A_\mu \\ \hline e\kappa\phi A_\nu & \phi \end{array}\right) \tag{25.1}$$

with $M, N = 0, 1, 2, 3, 5$: $\mu, \nu = 0, 1, 2, 3$, and the fields $g_{\mu\nu}$, A_μ, ϕ depending only on x^μ, the coordinates on M_4. The line element is

$$\mathrm{d}s_5^2 = \gamma_{MN}\mathrm{d}x^M\mathrm{d}x^N = \mathrm{d}s_4^2 + \phi(\mathrm{d}y + e\kappa A_\mu \mathrm{d}x^\mu)^2 \tag{25.2a}$$

where we used the notation $y \equiv x^5$ and

$$\mathrm{d}s_4^2 = g_{\mu\nu}\mathrm{d}x^\mu\mathrm{d}x^\nu \tag{25.2b}$$

is the line element on M_4. This $\mathrm{d}s_5^2$ admits a gauge invariance

$$\left.\begin{aligned} x^\mu &\to x'^\mu = x^\mu \\ y &\to y' = y + e\kappa\alpha(x^\mu) \\ A_\mu &\to A'_\mu = A_\mu - \partial_\mu\alpha. \end{aligned}\right\} \tag{25.3}$$

The ordinary gauge transformations of the vector field A_μ thus correspond to four-space-dependent translations on the circle. The gauge group acquires a five-dimensional geometrical meaning.

The five-dimensional Einstein–Hilbert action is

$$I_5 = -\frac{1}{16\pi G_5}\int \mathrm{d}x^5(|g_5|)^{1/2}R_5 \tag{25.4}$$

where the subscript 5, as in the line element earlier, always indicates that the corresponding quantities refer to five-space. Inserting here the metric (25.1) and carrying out the y integration yields

$$\left.\begin{aligned} I_4 &= \int (|g_4|)^{1/2}(|\phi|)^{1/2}\left(-\frac{1}{16\pi G}R_4 + \frac{e^2\kappa^2}{16\pi G}\phi g^{\mu\rho}g^{\nu\sigma}F_{\mu\nu}F_{\rho\sigma}\right) \\ G &= \frac{G_5}{2\pi\rho},\ g_4 = \det(g_{\mu\nu}),\ F_{\mu\nu} = \partial_\mu A_\nu - \partial_\nu A_\mu, \end{aligned}\right\} \tag{25.5}$$

where R_4 is the scalar curvature calculated from the four-metric $g_{\mu\nu}$, and ρ is the radius of the small circle in the fifth dimension. Were it not for the scalar field $\phi(x)$ this would be precisely the four-dimensional Einstein–Maxwell lagrangian. As is, this involves the Jordan–Brans–Dicke (JBD) modification thereof. The correct sign of the Maxwell piece requires the fifth dimension to be space-like, as does causality for that matter (already in flat space any closed curve in the plane defined by two time-like directions is everywhere time-like and violates causality.)

The basic set-up (Kaluza 1921, Klein 1926) does not change radically by introducing more than one, say N, extra space-like dimensions curled up into a small N-manifold M_N (de Witt 1964, Kerner 1968, Trautman 1970, Cho & Freund 1975). The Maxwell term in (25.5) is then replaced by a Yang–Mills term (with a nonabelian JBD modification), the gauge group being determined by the isometries of M_N. The scalar JBD fields self-interact as in a σ-model. The correct normalization of the Yang–Mills piece fixes the length

$$l \equiv 2\pi\kappa = \xi 4\pi\alpha^{-1/2}G \tag{25.6}$$

with $\alpha = e^2/4\pi$ the 'fine structure' constant of the Yang–Mills interaction presumably corresponding to some grand unification group, so that $\alpha \sim 10^{-2}$; $G = G_{4+N}/V_N$ (where V_N is the volume of M_N) is the four-dimensional Newton constant of gravity, ξ a parameter, $\xi = 1$ for a torus and in general $\xi > 1$ for coset spaces. (Wetterich 1983, Weinberg 1983). The length l is readily established as the 'size' of the small manifold (e.g., circumference for a circle). This size l as given by (25.6) exceeds the Planck length by 2–3 orders of magnitude thus calling into question the meaning of grand unification, since by the time the grand unification scale is reached space–time may cease to be four-dimensional. It is really remarkable that coming from the Planck scale via Kaluza–Klein arguments on the one hand, and from low energy physics via purely four-dimensional renormalization group arguments on the other hand, one ends up at the same scale in both cases. In Kaluza–Klein theories this means that quantum gravity is to be considered in the higher-dimensional (not the four-dimensional) context. Furthermore, this leads one to consider 'Kaluza–Klein cosmologies' in which the 'effective' number of space dimensions is time-dependent (Chodos & Detweiler 1980, Freund 1982, Alvarez & Belen Gavela 1983, Shafi & Wetterich 1983, Sahdev 1984).

In pure higher-dimensional gravity there is no reason for the vacuum to correspond to a product manifold, let alone to a product with one of the factors (the Lorentzian one) having dimension four. One could couple a judicious set of matter fields to the higher-dimensional gravity, in order to produce a classical solution of the type: four-dimensional Minkowski space $\times M_N$, with M_N a compact manifold, a phenomenon known as *spontaneous compactification* (Cremmer & Scherk 1976, Luciani 1978). But then one has to face the problem as to what determines the nature of the so postulated matter fields. Remarkably, in the interesting higher-dimensional supergravity (and superstring) theories, supersymmetry *requires* the presence of both Fermi and Bose matter as superpartners of the higher-dimensional graviton. As we shall see, the *Bose* superpartners of the graviton are capable of inducing spontaneous compactification with the unexpected result of preferentially leaving four dimensions large (Freund & Rubin 1980). We present the details in the next chapter.

26

Eleven-dimensional supergravity and its preferential compactification

We saw in the last chapter how an extremely simple theory in ten dimensions reduced to the famous finite (see chapter 18) $N=4$ Yang–Mills theory in four dimensions. Just as the $N=4$ Yang–Mills theory is maximal, in the sense of there not existing $N>4$ supersymmetric theories with supermultiplets containing only states of helicity between -1 and $+1$, so the ten-dimensional supersymmetric Yang–Mills theory is maximal. To see this, notice that the spectrum of a supersymmetric Yang–Mills theory in d-dimensions consists of one massless (i.e., transverse) vector, and one massless spinor both in the adjoint representation of the gauge group. For a gauge group of dimensionality γ we then have $\delta_B=\gamma(d-2)$ Bose degrees of freedom ($d-2$ since we subtract both the time-like and longitudinal modes of the vector) and $\delta_F=\gamma 2^{[\frac{1}{2}d]-\varepsilon_d}$ Fermi degrees of freedom. Here $[\frac{1}{2}d]$ is the largest integer in $\frac{1}{2}d$ (see chapter 3) and ε_d takes the values zero for Dirac, one for Weyl or Majorana and two for Majorana–Weyl spinors respectively. The existence of such spinors is specified in table 3.4. The equality

$$\delta_B=\delta_F \tag{26.1a}$$

required by supersymmetry (see chapter 5) thus takes the form

$$d-2=2^{[\frac{1}{2}d]-\varepsilon_d}. \tag{26.1b}$$

Reading ε_d from table 5.1 for Lorentzian metric signature, we see that (26.1) cannot be enforced beyond ten space–time dimensions; for ten-dimensions one must choose $\varepsilon_{10}=2$ corresponding to Majorana–Weyl spinors. What happens is, that δ_F increases exponentially with d, whereas δ_B only increases linearly with d and for $d>10$ the Bose degrees of freedom cannot 'keep pace' with the Fermi degrees of freedom. Of course there exist supermultiplets even for $d>10$, but they will of necessity include higher rank tensor and/or spin–tensor fields.

A similar argument can be made for supergravity, for which the maximal dimension turns out to be $d=11$ (Nahm 1978). The number of graviton degrees of freedom δ_g increases quadratically with d, $\delta_g=\frac{1}{2}d(d-3)$, corresponding to a symmetric, transverse traceless tensor of rank two. By contrast the spin-vector gravitino (transverse and obeying $\gamma^M \psi^\alpha_M = 0$) has $\delta_{\psi_M} = (d-3)2^{[\frac{1}{2}d]-\varepsilon_d}$ degrees of freedom. In eleven dimensions $\varepsilon_d = 1$, as there exist Majorana spinors, so that $\delta_{\psi_M} = 8(2)^{5-1} = 128$, whereas $\delta_g = \frac{1}{2}(11 \times 8) = 44$ and there arises a mismatch. The cure calls for a transverse antisymmetric tensor field of rank three with $\binom{d-2}{3}$ degrees of freedom, which in $d = 11$ yields precisely the 84 additional degrees of freedom to make supersymmetry possible: $44 + 84 = 128$, and $\delta_B = \delta_F$ as required by (26.1*a*). That this is the right assignment – as opposed, say, to introducing 84 scalars – can be checked either from superalgebra representation theory (Nahm 1978) or by explicitly constructing a supersymmetric lagrangian (Cremmer, Julia & Scherk 1978). As we said, this eleven-dimensional supergravity is maximal for Lorentzian metric signature. For two time and $(d-2)$ space dimensions there are Majorana–Weyl spinors for $d = 12$ (see table 3.4) and there may also exist a supergravity, though its interpretation would be quite obscure, anyway at present.

Continuing with the eleven-dimensional theory, we proceed to write down its lagrangian (Cremmer, Julia, & Scherk 1978) in component form:

$$\mathscr{L}_{11} = \mathscr{L}_B + \mathscr{L}_F \tag{26.2a}$$

$$\begin{aligned}\mathscr{L}_B = &-\tfrac{1}{2}eR - \tfrac{1}{48}eF_{MNPR}F^{MNPR}\\ &+\frac{\sqrt{2}}{3456}\varepsilon^{M_1\ldots M_{11}}F_{M_1\ldots M_4}F_{M_5\ldots M_8}A_{M_9\ldots M_{11}}\end{aligned} \tag{26.2b}$$

$$\begin{aligned}\mathscr{L}_F = &-e[\tfrac{1}{2}\bar{\psi}_M\Gamma^{MNP}D_N\psi_P + \frac{\sqrt{2}}{192}(\bar{\psi}_M\Gamma^{MNPQRS}\psi_N\\ &+ 12\bar{\psi}^P\Gamma^{QR}\psi^S)F_{PQRS}].\end{aligned} \tag{26.2c}$$

Here e^A_M and ω^{AB}_M are the *elfbein* and eleven-dimensional spin connection respectively, $e = \det e^A_M$, A_{MNR} is the antisymmetric tensor field required by super-symmetry, ψ^α_M the gravitino field,

$$\left.\begin{aligned}F_{MNPR} &= 24\partial_{[M}A_{NPR]}\\ \Gamma^{MN\ldots P} &= \gamma^{[M}\gamma^N\cdots\gamma^{P]} \quad (\text{in particular } \Gamma^M = \gamma^M),\end{aligned}\right\} \tag{26.2d}$$

the square bracket indicates the antisymmetrized sum over all permutations of bracketed indices, divided by the number of these permutations; γ^M are the real eleven-dimensional Majorana γ-matrices, and

$$\left.\begin{aligned}D_N\psi_M &= \partial_N\psi_M - \tfrac{1}{2}\omega^{AB}_N\sigma_{AB}\psi_M\\ \sigma_{AB} &= \tfrac{1}{2}\Gamma_{AB}.\end{aligned}\right\} \tag{26.2e}$$

We have omitted the four-Fermi terms. They can be included by appropriately 'supercovariantizing' the spin connection and the antisymmetric tensor field strengths (Cremmer, Julia & Scherk 1978). Our notations differ from those of this reference in our use of the $- + + \cdots +$ metric signature, of real γ-matrices, in the normalization of our A_{MNP} (which is $\frac{1}{6}$ times theirs) and in our units which set their gravitational constant $k^2 = \frac{1}{2}$ (see also Englert & Nicolai 1983).

The action $\int d^{11}x \mathscr{L}_{11}$ is invariant under the 'on shell' local supersymmetry transformation

$$\left.\begin{aligned} \delta e^A_M &= \tfrac{1}{2}\bar{\varepsilon}\Gamma^A\psi_M \\ \delta A_{MNP} &= -\frac{\sqrt{2}}{8}\bar{\varepsilon}\Gamma_{MN}\psi_P \\ \delta\psi_M &= D_M\varepsilon - \frac{\sqrt{2}}{288}(\Gamma^{NPQR}{}_M + 8\delta^N_M\Gamma^{PQR})F_{NPQR}\varepsilon. \end{aligned}\right\} \tag{26.3a}$$

Here also $\delta\psi_M$ acquires further Fermi bilinears through 'supercovariantization'. The action is also invariant under the gauge transformations

$$A_{MNR} \rightarrow A_{MNR} + \partial_M\Lambda_{NR} + \partial_N\Lambda_{RM} + \partial_R\Lambda_{MN} \tag{26.3b}$$

where $\Lambda_{MN}(x) = -\Lambda_{NM}(x)$ (the Chern–Simons term $\varepsilon^{M_1 \ldots M_{11}} F_{M_1 \ldots M_4} \times F_{M_5 \ldots M_8} A_{M_9 \ldots M_{11}}$ changes by an exact divergence under these transformations).

We note the simplicity of the lagrangian (26.2) and the absence of a cosmological term. The latter is connected with the nonexistence of a de-Sitter superalgebra in eleven-dimensional space–time (remember the arguments of chapters 21 and 22). After all, even in four dimensions, the existence of a de-Sitter superalgebra was intimately connected with the 'accidental' isomorphism $\mathscr{sp}(4) \sim \mathscr{so}(3, 2)$, and there are no such isomorphisms involving the eleven-dimensional de-Sitter or anti-de-Sitter algebras $\mathscr{so}(11, 1)$, $\mathscr{so}(10, 2)$.

With the eleven-dimensional supergravity lagrangian (26.2) now in hand (alas, its construction is not very methodical as yet, so we have skipped it), let us see whether there is any chance for spontaneous compactification of some of the ten space dimensions. There is 'Bose-matter' in this theory, given by the A_{MNR} field, and this does produce compactification. The surprise is that this compactification is preferential (Freund & Rubin 1980) towards a four-dimensional space–time.

To find possible vacua, we have to search for classical solutions around which to quantize the theory. There is of course the trivial vacuum

$$e_M^A = \delta_M^A, \quad \omega_M^{AB} = 0, \quad A_{MNR} = 0, \quad \psi_M^\alpha = 0, \tag{26.4}$$

a solution of the field equations following from the lagrangian (26.2). It corresponds to eleven-dimensional Minkowski space–time M_{11} or to a product manifold $M_d \times T_{11-d}(T_{11-d}$ being the compact $(11-d)$-dimensional torus). To find less trivial solutions, let us first write down the field equations, for the special case $\psi_M^\alpha = 0$ (as Fermi fields should vanish in the vacuum).

$$\left.\begin{aligned} R_{MN} - \tfrac{1}{2}g_{MN}R &= -\tfrac{1}{6}(F_{MPQR}F_N{}^{PQR} - \tfrac{1}{8}g_{MN}F_{PQRS}F^{PQRS}) \\ D_M F^{MNPQ} &= -\frac{\sqrt{2}}{1152}\varepsilon^{NPQR_1\ldots R_8}F_{R_1\ldots R_4}F_{R_5\ldots R_8} \\ g_{MN} &= e_M^A e_N^B \eta_{AB}. \end{aligned}\right\} \tag{26.5}$$

We want the 'ordinary' space–time to be maximally symmetric. The gauge invariant antisymmetric field-strengths tensor F^{MNRS} can then have a non-vanishing vacuum expectation value if M_{11} is of the form $M_{11} = M_4 \times M_7$. The metric on M_{11} is then of the form

$$g_{MN} = \left(\begin{array}{c|c} g_{\mu\nu}(x) & 0 \\ \hline 0 & g_{mn}(y) \end{array}\right) \tag{26.6a}$$

where $M, N = 0, 1, 2, 3, 5, 6, \ldots, 11$; $\mu, \nu = 0, \ldots, 3$; $m, n = 5, \ldots, 11$, $x^\mu(y^m)$ are coordinates on $M_4(M_7)$. The F^{MNRS} can then (Freund & Rubin 1980) take the form (this is not the most general form possible)

$$F^{MNRS} = \begin{cases} \dfrac{\varepsilon^{MNRS}}{(|g_4|)^{1/2}}f & \text{when } M, N, R, S \text{ all take values between 0 and 3} \\ 0 & \text{otherwise.} \end{cases} \tag{26.6b}$$

Here ε^{MNRS} is the usual four-dimensional totally antisymmetric Levi–Civita symbol and g_4 is the determinant of the 4×4 matrix $g_{\mu\nu}(x)$ of equation (26.6*a*). For this vacuum field ansatz the Fermi field again vanishes.

$$\psi_M^\alpha = 0. \tag{26.6c}$$

Inserting the ansatz (26.6) into the A_{MNP} field equations we find

$$\frac{1}{(|g_4 g_7|)^{1/2}}\partial_M\left[(|g_4 g_7|)^{1/2}\frac{\varepsilon^{MNRS}}{(|g_4|)^{1/2}}f\right] = 0. \tag{26.7a}$$

From (26.6*b*) the index M in equation (26.7*a*) must take values between 0 and 3, so that ∂_M is necessarily of the form $\partial/\partial x^\mu$. Since g_7 depends only on y, and ε^{MNRS} is constant, this means

$$\frac{\partial}{\partial x^\mu} f = 0 \tag{26.7b}$$

The Bianchi identities

$$\frac{1}{(|g_4 g_7|)^{1/2}} \partial_M [(|g_4 g_7|)^{1/2} \varepsilon^{MM_2 \ldots M_{11}} F_{M_8 \ldots M_{11}}] = 0 \tag{26.7c}$$

similarly yield

$$\frac{\partial}{\partial y^m} f = 0 \tag{26.7d}$$

so that from (26.7*b*) and (26.7*d*) we find

$$f = \text{constant} \tag{26.8}$$

Inserting this result into the expression of the energy–momentum tensor on the right-hand side of the first equation (26.5), we see that the second term of this tensor gives an overall cosmological term, whereas the first term gives an extra cosmological term whenever both M and N are between 0 and 3. In other words M_4 and M_7 are both Einstein manifolds but corresponding to different values Λ_4 and Λ_7 of the cosmological constant. We have

$$\Lambda_4 = -\tfrac{1}{4} g^{\mu\nu} R_{\mu\nu} = \tfrac{4}{6} f^2 \operatorname{sgn}(g_4), \quad \Lambda_7 = -\tfrac{5}{14} g^{mn} R_{mn} = -\tfrac{5}{6} f^2 \operatorname{sgn}(g_4) \tag{26.9}$$

The sign of g_4 enters since the ε symbol in the ansatz (26.6*b*), contracted with the metric tensors in the energy–momentum tensor yields a determinant of $g_{\mu\nu}$, whereas the factor $1/\sqrt{|g_4|}$ in the ansatz (26.6*b*), when squared (the energy–momentum tensor is quadratic in F_{MNPQ}) yields a factor $1/|g_4|$, so that we find a factor $g_4/|g_4| = \operatorname{sgn}(g_4)$. A positive value of the cosmological constant means compactification. f being real, f^2 is positive so that the signs of Λ_4 and Λ_7 depend solely on the sign of g_4, i.e., on whether the time-like dimension is on $M_4 (g_4 < 0)$ or on $M_7 (g_4 > 0)$. We see from (26.9) that (i) when the time-like dimension is on M_4, then M_7 compactifies, and (ii) when it is on M_7 then M_4 compactifies. Case (i) is of course 'realistic'. We have thus shown that spontaneous compactification does occur in eleven dimensional supergravity. That already is interesting, but we also found that there exists a *preferential* compactification towards four space–time dimensions. That is a surprise. Let us briefly recapitulate the logic of this argument. Eleven is the maximum dimensionality of a space–time in which a supergravity exists. Supersymmetry *requires*, in addition to the graviton, the existence of a further Bose matter-field represented by an antisymmetric tensor A_{MNP} with three indices. That this tensor has three indices is thus a consequence of supersymmetry. Its gauge invariant curl then has four indices and it is this curl that can have a vacuum expectation value in a, then necessarily, four-dimensional space. If time is among the four dimensions then the dynamics is right to compactify the remaining seven dimensions. It is now clear that this preferential $4+7$ split of the originally eleven-dimensional space–time can be traced directly to supersymmetry. It is thus supersymmetry that 'dials' the dimensionality of the observed space–time.

Yet as it stands, this argument still has some weak spots and outright difficulties. First of all the time dimension could be among the seven rather than the four. No good argument exists to rule out this alternative. Even if time is on M_4, this M_4 is not then a Minkowski space with vanishing cosmological constant, but rather an anti-de-Sitter space with an immense (four-dimensional) cosmological constant given by (26.9). Grand unification is also plagued by outsize cosmological constants, and understanding why the observed Λ_4 is so very close to zero is a major theoretical puzzle, whether for grand unification or Kaluza–Klein theory. There is in Kaluza–Klein theory the additional deficiency (Witten 1985) that it is hard to obtain chiral fermions at the four-dimensional level. This may be somewhat alleviated in supergravity, when bound states and 'hidden' symmetries like the Cremmer–Julia $SU(8)$ of chapter 24 are taken into account.

For the time being we ignore these problems and inquire into the type of physics we should expect in four dimensions. This depends largely on the 'shape' of the small Einstein manifold M_7. The isometry group G_7 (the invariance group of the metric) on M_7 dictates the overt gauged symmetry of the four-dimensional theory (in addition to this, one has the gauged hidden symmetry). So we will have a gauged $Sp(4)$, corresponding to gravity with cosmological term, on M_4 and a gauged G_7. The Lie algebra $\mathfrak{sp}(4) \oplus \mathfrak{g}_7$ is then spanned by the Killing vector fields on $M_4 \times M_7$. But even some supersymmetry can survive down to four dimensions. If this is the case, then the vanishing of the Fermi field ψ^α_M (equation (26.6*c*)) must be maintained under some supersymmetry transformations. But the change of ψ^α_M under a supersymmetry transformation as given by equation (26.3*a*) must then vanish.

$$\delta\psi^\alpha_M = 0. \tag{26.10}$$

The ε^α appearing in equations (26.3) are 32-component Majorana spinors. Now assume the factorization

$$\varepsilon(x, y) = \varepsilon(x)\eta(y) \tag{26.11}$$

with ε a four-component. Majorana spinor on M_4, and η an eight-component spinor on M_7. Equation (26.10) then requires

$$\hat{D}_m\eta \equiv \left(D_m - \mathrm{i}\frac{f}{6\sqrt{2}}\gamma_m\right)\eta = 0 \tag{26.12}$$

where $D_m\eta$ is the ordinary covariant derivative on M_7 of the spinor η. The Killing spinor equations (26.12) have at most eight independent solutions (η has eight components). So, by equation (26.11) we can then have at most $N = 8$ supersymmetry in four dimensions, as expected. The integrability condition of equation (26.12) is (Awada, Duff and Pope 1983)

$$[\hat{D}_m, \hat{D}_n]\eta = C_{mnab}[\gamma^a, \gamma^b]_-\eta = 0 \tag{26.13}$$

where m, $n(a, b)$ are world (tangent space) indices on M_7 and C_{mnab} is Weyl's conformal curvature tensor. The $[\gamma^a, \gamma^b]_-$ span the Lie algebra of $\mathfrak{spin}(7)$, whereas the combinations $C_{mnab}[\gamma^a, \gamma^b]_-$ span a subalgebra w thereof: the Weyl holonomy algebra. According to equation (26.13) η must be a singlet of this Weyl holonomy algebra w. There will then be at most as many Majorana-spinor supersymmetry generators in four dimensions as there are w-singlets in the spinorial octet representation of $\mathfrak{spin}(7)$.

We can now go to specific cases. The simplest case is to have M_7 conformally flat in which case equation (26.13) reduces to $0 = 0$ and there are no conditions on η so that we have the maximum number of eight independent Majorana supersymmetries in four-dimensions. This can happen only in two cases:

(i) for the seven-torus T^7 (26.4) corresponding to the Cremmer–Julia $N = 8$ supergravity in four-dimenssions and

(ii) for the seven-sphere S^7, with isometry group $SO(8)$ corresponding to the de Wit–Nicolai version of $N = 8$ supergravity (see chapter 24).

But there are many additional solutions with $N = 1$, 2, 3, 4 or without supersymmetry ($N = 0$), corresponding to choices of the various Einstein manifolds M_7 (see Awada Duff & Pope 1983, Castellani, d'Auria & Fré 1984). In the T^7 and S^7 cases just mentioned, we of course find also the hidden gauged $SU(8)$ symmetry. In addition to the massless supergravity sector in this eleven-dimensional theory we also have the massive higher harmonics. The full theory is thus *not* the same as its massless supergravity truncation. In fact, in the case of S^7 there exist massless (by anti-de-Sitter standards) scalar states in addition to those 70 included in the four-dimensional $N=8$ supergravity multiplet. These do not have massless superpartners. The supersymmetry, after all, in $OSp(8|4)$ and not the $N=8$ Poincaré supergroup. Therefore the mass squared operator does *not* commute with the $OSp(8|4)$ Fermi generators so that states of different mass can belong to the same supermultiplet.

Next we recall that the ansatz (26.6*b*) is not the most general one. One can extend it without losing the maximal symmetry of M_4, by allowing $F_{MNPQ} \neq 0$ for $M, N, P, Q = 5, \ldots, 11$ (Englert 1982). This leads to further solutions with a remarkable geometric *raison d'être*. Furthermore if M_7 is not a homogeneous space one can also relax the ansatz (26.6*a*) by multiplying $g_{\mu\nu}(x)$ in the upper left-hand corner by an overall function $h(y)$ of the coordinates on M_7 (de Wit & Nicolai 1984, van Nieuwenhuizen 1984). In this case the compact manifold M_7 need not even be Einstein. These solutions permit one to find more simply, coming from eleven dimensions, all the extrema of the scalar potential of the four-dimensional $N=8$ theory (Warner 1984). This is not to say that to each and every $M_4 \times M_7$ solution of the eleven-dimensional theory there corresponds an extremum of the four-dimensional scalar potential. For instance, the case of the 'squashed' seven-sphere (Awada, Duff & Pope 1983, Duff, Nilsson & Pope 1983), does not correspond to such an extremum. The reason is that in the process of squashing the ordinary seven-sphere a level crossing occurs: previously very massive higher harmonics descend to zero mass, whereas previously massless levels acquire immense mass. Since the higher harmonics of the Kaluza–Klein theory compactified on S^7 are not present in ordinary four-dimensional $N=8$ supergravity, the symmetry breaking corresponding to sphere squashing can not be replicated in four-dimensional $N=8$ supergravity. The massive states, a signature of higher dimensions, play a crucial role. This phenomenon is referred to as *space invaders*.

The next question concerns the stability of these solutions under classical perturbations. Where the full spectrum is known (e.g., $M_7 = S^7$) stability amounts to the requirement that no tachyons be present in the spectrum. For the $N=8$ supersymmetric $M_7 = S^7$ case, there are indeed no tachyons in the spectrum (Englert & Nicolai 1983), so the solution (26.6), (26.8) with $M_7 = S^7$ is classically stable. So are all solutions that have any ($N \geqslant 1$) residual supersymmetry in four dimensions. Of the nonsupersymmetric solutions both stable and unstable ones are known. For instance, all Englert solutions (without the de Wit–Nicolai $g_{\mu\nu}(x)h(y)$ generalization mentioned above) have no residual supersymmetry and are unstable.

As mentioned above, there is the problem of the large four-dimensional cosmological constant Λ_4. Ideally, we would wish $\Lambda_4 = 0$, or at least a cosmological (i.e., time-dependent rather than static) solution that yields a small Λ_4 at later times. Cosmological solutions to eleven-dimensional supergravity have by now been found (Freund 1982).

For the higher-dimensional theory to make sense at the quantum level, it must be a finite quantum field theory, for there are no renormalizable local field theories in space–times of dimension seven and higher, and anyway any local field theory which includes gravity is fated to be nonrenormalizable. At the one-loop level eleven-dimensional supergravity is finite, but this is trivial for an odd-dimensional space–time (Duff & Toms 1982). To have a nontrivial test one would have to check two loop finiteness. This has not yet been done for eleven-dimensional supergravity, but there exist possible dangerous counterterms, so the prospects are not the best. Still, only a calculation can settle this problem.

The potential divergence of eleven-dimensional supergravity has led to a retreat to ten space–time dimensions. In ten dimensions there exist two distinct $N = 2$ supergravities and an $N = 1$ supergravity which can be coupled to a $N = 1$ supersymmetric Yang–Mills system for some gauge group G. All these theories are limits for infinite string tension of supersymmetric string theories (Ramond 1971, Neveu & Schwarz 1971, Green & Schwarz 1982). The superstring theories are demonstrably one-loop finite (in ten dimensions this *is* nontrivial) and there exist good reasons to expect their finiteness upon inclusion of all higher loops as well. The consistency of the $N = 1$ theories requires the gauge group to be $SO(32)$ or $E_8 \times E_8$ (Green & Schwarz 1984*a*, Freund 1985, Gross, Harvey, Martinec & Rohm 1985). The $E_8 \times E_8$ case has phenomenological merits (Candelas, Horowitz, Strominger & Witten 1985). It should also be pointed out that these superstring theories do yield chiral fermions upon compactification down to four dimensions.

Roughly speaking, one-loop finiteness is achieved in superstring theory in a manner reminiscent of the way Glashow–Weinberg–Salam theory achieves the renormalizability absent in the old phenomenological Fermi theory of weak interactions (Green & Schwarz 1984). There, the offending four-Fermi vertices are smeared out by massive intermediate bosons, the Ws and the Z^0. In superstring theory, the infinitely many offensive vertices of supergravity theory (which like all local theories that include gravity is nonpolynomial) are smeared out by supermassive string excitations. If this finiteness is to hold up at higher loops, as seems now virtually certain, then a combination of supersymmetry, of the generalized Kaluza–Klein idea, and of the Veneziano–Nambu string idea may finally have achieved the long sought-after synthesis of quantum theory and general relativity. At the same time supersymmetry will have returned to string theory where it was originally conceived. All this would be yet another example of a phenomenon often encountered in physics: a theory, originally pursued for its mathematical and philosophical beauty, is shown to possess a new feature which then turns into the driving force of research into the theory. Renormalizability ('t Hooft 1971) was this new feature for electroweak theory, finiteness in the presence of gravity (?) may just be it for supergravity and superstring type theories.

Part IV
Conclusion

27

The present status of supersymmetry

It may be appropriate to conclude this book by assessing the present status of the principle of supersymmetry in physics. A physical principle can be reliably evaluated according to the following criteria:

(A) The experimental evidence that supports the principle.
(B) New phenomena predicted on the basis of the principle.
(C) The experimental and theoretical puzzles solved by the principle.
(D) The internal consistency of theories that incorporate the principle.
(E) The aesthetic and philosophic advantages of the principle.

Let us now consider each of these five criteria and apply them to supersymmetry.

As the first three parts of this book imply, supersymmetry fares poorly on criterion (A). There is no hard evidence for supersymmetry in particle physics (at the time of this writing). Given the mathematical novelty of the concept, one may wonder whether supersymmetric systems appear in physics at all. There are two fields of physics in which supersymmetry does indeed make an appearance: the quantum statistical mechanics of two-dimensional systems and nuclear physics. We briefly review these examples with the main objective of showing that the set of supersymmetric physical systems is not empty.

First the example from the theory of two-dimensional systems. The tricritical Ising model, realized experimentally by adsorbing helium-4 on krypton plated graphite (notice the *de rigueur* appearance of krypton) near the critical point is supersymmetric. Near a critical point two-dimensional statistical systems scale and become conformally invariant. In two dimensions the conformal algebra is the infinite-dimensional Virasoro algebra corresponding to conformal mappings. This algebra allows a supersymmetric extension, as known already from the theory of dual models (Ramond 1971, Neveu & Schwarz 1971). It is this infinite-dimensional superalgebra that is the supersymmetry algebra of the tricritical Ising model. Its validity implies 'selection rules' and 'Clebsch–Gordan'-like relations, which are found to hold (Friedan, Qiu & Shenker 1985). At the critical point, this model is exactly soluble, so these supersymmetry relations are only partial features of the exact solution. It is as if, with the intent of showing the relevance of ordinary Lie algebras beyond *so*(3) for physics, we were to parade the exactly soluble hydrogen-atom with its *so*(4) invariance. Sure, the degeneracy of states with equal principal quantum number *n*, is evidence for this 'accidental' symmetry, but if this were the only system in nature with symmetry beyond *so*(3), ordinary Lie algebras would play a less dominant role in physics than they now do. In the same spirit the tricritical Ising model settles the existence question, but can hardly be used to justify the considerable body of work on supersymmetry over the last decade.

The nuclear physics example (Balantekin, Bars & Iachello 1981) involves only an approximate supersymmetry and concerns itself with the details of the supersymmetry breaking. For nuclei in the osmium–platinum region a model involving bosonic spin singlet-like nucleon pairs of total angular momentum $j=0$ and $j=2$, and fermionic $^2D_{3/2}$ (i.e., $J=\frac{3}{2}$) unpaired protons is proposed. This involves six Bose degrees of freedom (one for the $j=0$, and five for the $j=2$ pairs) and four Fermi degrees of freedom (corresponding to $j=\frac{3}{2}$). Now, disregarding the difference between single nucleons and nucleon pairs (!), one can postulate a unitary $U(6|4)$ supersymmetry between these $10=6+4$ degrees of freedom. By its very nature, such a supersymmetry must be broken. There are various 'chains' for this supersymmetry breaking, and to each chain there corresponds a mass formula involving a linear combination (with free coefficients) of the Casimir operators of the (super) groups in the breakdown chain. The most successful chain is

$$U(6|4)\to U(6)\times U(4)\to SO(6)\times SU(4)\to Spin(6)\to Spin(5)$$
$$\to Spin(3)\to Spin(2).$$

The corresponding ten parameter mass formula can fit many levels of both even–even and even–odd nuclei at the 20–30% level of accuracy. Again, while certainly very interesting, this is not the kind of phenomenon to make supersymmetry physically compelling. A similar model in hadron phenomenology was put forward much earlier by Miyazawa (1968), and developed recently by Gürsey (1984).

All told, criterion (*A*), as applied to supersymmetry today is far from providing clear evidence in favor of supersymmetry. Given this bleak assessment, one might be tempted to conclude that criteria (*B*) and (*C*) will also fail to help make the case for supersymmetry. This is not so! We have seen in chapter 17 that supersymmetric grand unification makes predictions concerning supersymmetric partners to known particles.

These theories can be tested in the near future, hopefully with results that will change the verdict according to criterion (*A*). Similar considerations apply to extended supergravities in four-dimensions to higher-dimensional supergravities (chapters 23, 26) and to superstrings. So on criterion (*B*) the prospects are brighter.

On criterion (*C*) supersymmetry has not solved any experimental puzzles, partly because for a long time there just hadn't been any such puzzles around! Concerning theoretical puzzles, the hierarchy problem is certainly an important and venerable puzzle going back essentially to Dirac's large number hypothesis (Dirac 1937). Here as we saw in chapter 17, supersymmetry is of use. Further problems such as why space–time appears four-dimensional, and the force/matter problem also receive a first meaningful treatment in the framework of supersymmetry. Based on criterion (*C*) then, supersymmetry does make an impact.

On criterion (*D*) supersymmetry fairs very well indeed. After all, who would have thought that virtually half a century into quantum field theory, we could still face the surprise of finite theories. There is no question that such theories have a degree of consistency beyond that of ordinary renormalizable theories. Even gravity, as we saw in chapter 26, appears to be consistently quantized this way.

On criterion (*E*) supersymmetry is also a clear winner. New mathematical structures have emerged: superalgebras, supergroups, supermanifolds. A much more unified picture is possible, fermions and bosons are meaningfully grouped together, and again we can refer to the resolution of the age old force/matter problem. It is after all to a large extent these aesthetic, mathematical and philosophical merits that have driven much of the early work on supersymmetry. In a sense I perceive a certain similarity with the history of nonabelian gauge theories. They also were formulated by Yang and Mills for aesthetical, mathematical and philosophical reasons, rather than brought forth in answer to existing experimental puzzles. Proposed in the mid-fifties, they entered the mainstream of physics only some 17 years later. Why this long delay? This question has a fascinating answer which may have some relevance for supersymmetry.

In the early days of Yang–Mills theory it was believed (Yang & Mills 1954, Sakurai 1960) that 'flavor'-symmetries, such as isospin, baryon number, hypercharge, 'eightfold way $SU(3)$', charm, were to be gauged. The corresponding gauge bosons would be massless vector mesons with quantum numbers like the ρ, ω, ϕ, K^*, J/ψ etc.... No mesons with these attributes had yet been discovered at the time this proposal was particularly forcefully put forward by Sakurai. It was clear that these mesons had to 'somehow' acquire mass. So, as the ρ, ω, ϕ, K^* mesons were discovered everybody seemed to agree that Sakurai's predictions were being confirmed. To make matters 'worse', the coupling pattern of these vector mesons was following the rules one would have expected for gauge bosons (e.g., the ρ^0, which was to couple to the third component I_3 of isospin, did couple half as strongly to the proton ($I_3 = \frac{1}{2}$) as it did to the π^+ meson ($I_3 = +1$)). Gauge theory had arrived; all that remained to be explained was the nonvanishing mass of these flavor gauge bosons. Imaginative people came up with mechanisms by which gauge bosons grow mass (Englert & Brout 1964, Higgs 1964, Guralnik, Hagen & Kibble 1964, Schwinger 1962). Alas, these mechanisms didn't work! According to the Brout–Englert–Higgs mechanism, the K^* mesons acquired mass as $SU(3)_{\text{eightfold way}}$ was being broken, i.e., at the level of strong interactions, whereas the $\rho^\pm$ mesons only acquired their mass when isospin was broken, i.e., at the level of electromagnetic interactions. One then expected $(m_{\rho^\pm}/m_{K^*})^2 \sim 1/100$ whereas experimentally it is close to one (Schwinger's model referred to two space–time dimensions, and its relevance to the mass problem in four dimensions was unclear at the time). What did one (we all?) conclude from this? Well, since we 'knew' Sakurai's theory to work, too bad for Brout, Englert, Higgs! This way a remarkable theoretical idea was shelved for about six years. We now know (or 'know'?) the gauged symmetry to be not $SU(3)_{\text{eightfold way}}$, but rather $SU(3)_{\text{color}} \times (SU(2) \times U(1))_{\text{electroweak}}$, and here the Higgs mechanism does work. The ρ, K^*, *et al.*, have long been demoted to 'mere' quark–antiquark $^3S_1$ bound states. Their coupling pattern has been successfully explained by the quark model and the successful vector dominance hypothesis (originally due to Nambu 1957, Gell-Mann & Zachariasen 1961, later extended by Freund 1966, and by Ross & Stodolsky 1966). How about Sakurai's theory? We now know it to be obsolete. Yet without it, we might not have had (*A*) the eightfold way[†], (*B*) vector-dominance, (*C*) the Brout–Englert–Higgs mechanism, and (*D*) the Schwinger model. We can now conveniently discard[††] the flawed parent theory, and gracefully retain its four remarkable offspring. The Sakurai theory has thus played both a positive role by leading to these four successful ideas, and a negative role by retarding their acceptance.

[†] Both the Gell-Mann and the Ne'eman papers (Gell-Mann & Ne'eman 1964) are formulated in the context of Sakurai's ideas.

[††] Ironically, it has been proposed very recently that the ρ, ω, ϕ, K^* might also be viewed as dynamical (composite) gauge bosons of a hidden $U(3)$ symmetry in a nonlinear chiral effective lagrangian for QCD (Bando, Kugo, Uehara, Yamawaki & Yanagida 1985), thus vindicating Sakurai's intuition to a certain extent.

I have allowed myself this historical detour because I believe it to hold a moral for the present status of supersymmetry. It may well be that we already possess all the essential ingredients for a successful implementation of supersymmetry in particle physics, but are blinded by an assumption as obvious to 'us all', as Sakurai's ideas about flavor gauging were in their time. One of the hardest steps in science is the discarding of 'obvious' flaws. Let me not speculate in writing about what the possible obvious flaws in present-day supersymmetry ideas may be. After all, there is always the obvious way out: not enough accelerator energy. Time will tell.

References

Abbott, L.F., Grisaru, M.T. & Schnitzer, H.J. (1977), Cancellation of the Supercurrent Anomaly in a Supersymmetric Gauge Theory, *Phys. Lett.* **71B**, 161–4.

Adler, S. & Bardeen, W. (1969), Absence of Higher Order Corrections in the Anomalous Axial Vector Divergence Equation, *Phys. Rev.* **182**, 1517–36.

Alvarez, E. & Belen Gavela, M. (1983), Entropy from Extra Dimensions, *Phys. Rev. Lett.* **51**, 931–4.

Alvarez-Gaumé, L. & Freedman, D.Z. (1981), Geometrical Structure and Ultraviolet Finiteness in the Supersymmetric σ-model, *Comm. Math. Phys.* **80**, 443–51.

Appelquist, T., Chodos, A. & Freund, P.G.O. (1986), *Modern Kaluza–Klein Theory* (Benjamin/Cummings, Reading, Mass.) to appear.

Aragone, C. & Deser, S. (1979), Consistency Problems of Hypergravity, *Phys. Lett.* **86B**, 161–3.

Arnowitt, R. & Nath, P. (1976), Riemannian Geometry in Spaces with Grassmann Coordinates, *Gen. Rel. Grav.* **7**, 89–103.

Atiyah, M., Bott, R. & Shapiro, A., (1964), Clifford Modules, *Topology* **3** (suppl. 1), 3–38.

Avdeev, L.V., Tarasov, O.V. & Vladimirov, A.A. (1980), Vanishing of the Three-Loop Charge Renormalization Function in a Supersymmetric Gauge Theory, *Phys. Lett.* **96B**, 94–6.

Awada, M.A., Duff, M.J. & Pope, C.N. (1983), $N = 8$ Supergravity Breaks Down to $N = 1$, *Phys. Rev. Lett.* **50**, 294–7.

Bagger, J. & Witten, E. (1983), Matter Couplings in $N = 2$ Supergravity, *Nucl. Phys.* **B222**, 1–10.

Balantekin, A.B., Bars, I. & Iachello, F. (1981), $U(6|4)$ Supersymmetry in Nuclei, *Nucl. Phys.* **A370**, 284–316.

Bando, M., Kugo, T., Uehara, S., Yamawaki, K. & Yanagida, T. (1985), Is the ρ Meson a Dynamical Gauge Boson of Hidden Local Symmetry?, *Phys. Rev. Lett.* **54**, 1215–18.

Bargmann, V., Michel, L. & Telegdi, V.L. (1959), Precession of the Polarization of Particles Moving in a Homogeneous Electromagnetic Field, *Phys. Rev. Lett.* **2**, 435–6.

Bargmann, V. & Wigner, E.P. (1948), Group Theoretical Discussion of Relativistic Wave Equations, *Proc. Nat. Acad. Sci. USA.* **34**, 211–23.

Batchelor, M. (1980), Two Approaches to Supermanifolds, *Trans. Am. Math. Soc.* **258**, 257–70.

Berends, F.A., van Holten, J.W., de Wit, B., & van Nieuwenhuizen, P. (1980), On Spin $-5/2$ Gauge Fields, *J. Phys.* **A13**, 1643–9.

Berezin, F.A. (1966), *The Method of Second Quantization* (Academic Press, NY).

Berezin, F.A. (1979), Differential Forms on Supermanifolds, *Soviet J. Nucl. Phys.* **30**, 605–9.
Berezin, F.A. & Marinov, M.S. (1977), Particle Spin Dynamics as the Grassmann Variant of Classical Mechanics, *Ann. Phys.* (*NY*) **104**, 336–62.
Berg, B., Karowski, M. & Thun, H.J. (1976), Conserved Currents in the Massive Thirring Model, *Phys. Lett.* **64B**, 286–8.
Bishop, R.L. & Crittenden, R.J. (1964), *Geometry of Manifolds* (Academic Press, NY).
Breitenlohner, P. (1977), A Geometric Interpretation of Local Supersymmetry, *Phys. Lett.* **67B**, 49–51.
Brink, L., Deser, S., Zumino, B., di Vecchia, P. & Howe, P. (1976), Local Supersymmetry for Spinning Particles, *Phys. Lett.* **64B**, 435–8.
Brink, L., Gell-Mann, M., Ramond, P. & Schwarz, J.H. (1978), Extended Supergravity as Geometry of Superspace, *Phys. Lett.* **76B**, 417–22.
Brink, L., Lindgren, O. & Nilsson, B.E.W. (1983), The Ultraviolet Finiteness of the $N = 4$ Yang–Mills Theory, *Phys. Lett.* **123B**, 323–8.
Brink, L., Scherk, J. & Schwarz, J.H. (1977), Supersymmetric Yang–Mills Theories, *Nucl. Phys.* **B121**, 77–92.
Brink, L. & Schwarz, J.H. (1981), Quantum Superspace, *Phys. Lett.* **100B**, 310–12.
Calabi, E. (1979), Metriques Kählеriennes et Fibrés Holomorphes, *Ann. Sci. Ecole Norm. Sup.* **12**, 269–94.
Candelas, P., Horowitz, G., Strominger, A. & Witten, E. (1985), Vacuum Configurations for Superstrings, *Nucl. Phys.* **B258**, 46–74.
Candlin, D.J. (1956), On Sums over Trajectories for Systems with Fermi Statistics, *Nuovo Cim.* (10) **4**, 231–9.
Carruthers, P. (1971), Broken Scale Invariance in Particle Physics, *Phys. Rep.* **1C**, 1–30.
Cartan, E. (1894), Sur la Structure des Groupes de Transformations Finis et Continuus, Thèse, Paris.
Casalbuoni, R. (1976), On the Quantization of Systems with Anticomuting Variables, *Nuovo Cim.* **33A**, 115–25.
Casalbuoni, R. (1976*a*), The Classical Mechanics, for Bose–Fermi Systems, *Nuovo Cim.* **33A**, 389–431.
Castellani, L., d'Auria, R. & Fré, P. (1984), $SU(3) \times SU(2) \times U(1)$ from $D = 11$ Supergravity, *Nucl. Phys.* **B239**, 610–52.
Caswell, W.E. & Zanon, D. (1981), Vanishing Three-Loop Beta Function in $N = 4$ Supersymmetric Yang–Mills Theory, *Phys. Lett.* **100B**, 152–6.
Chamseddine, A.H. & West, P.C. (1977), Supergravity as a Gauge Theory of Supersymmetry, *Nucl. Phys.* **B129**, 39–44.
Chevalley, C. (1954), *The Algebraic Theory of Spinors* (Columbia University Press, NY).
Cho, Y.-M. & Freund, P.G.O. (1975), Non-Abelian Gauge Fields as Nambu–Goldstone Fields, *Phys. Rev.* **D12**, 1711–20.
Chodos, A. & Detweiler, S. (1980), Where has the Fifth Dimension gone? *Phys. Rev.* **D21**, 2167–70.
Clark, T., Piguet, O. & Sibold, K. (1978), Supercurrent, Renormalization and Anomalies, *Nucl. Phys.* **B143**, 445–84.
Coleman, S. & Mandula, J. (1967), All Possible Symmetries of the S Matrix, *Phys. Rev.* **159**, 1251–6.
Coquereaux, R. (1982), Modulo 8 Periodicity of Real Clifford Algebras and Particle Physics, *Phys. Lett.* **115B**, 389–95.

Corwin, L., Ne'eman, Y. & Sternberg, S. (1975), Graded Lie Algebras in Mathematics and Physics, *Rev. Mod. Phys.* **47**, 573–603.
Cremmer, E., Ferrara, S., Girardello, L. & van Proeyen, A. (1983), Yang–Mills Theories with Local Supersymmetry: Lagrangian Transformation Laws and Super-Higgs Effect, *Nucl. Phys.* **B212**, 413–42.
Cremmer, E. & Julia, B. (1979), *SO*(8) Supergravity, *Nucl. Phys.* **B159**, 141–212.
Cremmer, E., Julia, B. & Scherk, J. (1978), Supergravity Theory in 11 Dimensions, *Phys. Lett.* **76B**, 409–12.
Cremmer, E., Julia, B., Scherk, J., Ferrara, S., Girardello, L., & van Nieuwenhuizen, P. (1979), Spontaneous Symmetry Breaking and Higgs Effect in Supergravity without Cosmological Constant, *Nucl. Phys.* **B147**, 105–31.
Cremmer, E. & Scherk, J. (1976), Spontaneous Compactification of Space in an Einstein–Yang–Mills–Higgs Model, *Nucl. Phys.* **B108**, 409–16.
Curtright, T.L. (1977), Conformal Spinor Current Anomalies, *Phys. Lett.* **71B**, 185–8.
Curtright, T.L. (1979), Massless Field Supermultiplets with Arbitrary Spin, *Phys. Lett.* **85B**, 219–24.
Curtright, T.L. (1984), Private communication.
Curtright, T.L. & Freedman, D.Z. (1979), Nonlinear Sigma-Models with Extended Supersymmetry in Four Dimensions *Phys. Lett.* **90B**, 71; Erratum **91B**, 487.
Curtright, T.L. & Freund, P.G.O. (1979), *SU*(8) Unification and Supergravity, in *Proc. Stony Brook Supergravity Workshop*, P. van Nieuwenhuizen, & D.Z. Freedman, editors (North-Holland, Amsterdam) pp. 197–201.
Das, A. & Freedman, D.Z. (1977), Gauge Internal Symmetry in Extended Supergravity, *Nucl. Phys.* **B120**, 221–30.
Dell, J. & Smolin, L. (1979), Graded Manifold Theory as the Geometry of Supersymmetry, *Comm. Math. Phys.* **66**, 197–221.
Deser, S. & Zumino, B. (1976), Consistent Supergravity, *Phys. Lett.* **62B**, 335–7.
Deser, S. & Zumino, B. (1977), Broken Supersymmetry and Supergravity, *Phys. Rev. Lett.* **38**, 1433–6.
de Wit, B. & Nicolai, H. (1982), $N = 8$ Supergravity with local $SO(8) \times SU(8)$ Invariance, *Phys. Lett.* **108B**, 285–90.
de Wit, B. & Nicolai, H. (1984), A New *SO*(7) Invariant Solution of $d = 11$ Supergravity, *Phys. Lett.* **148B**, 60–4.
de Witt, B.S. (1964), in *Relativity, Groups and Topology*, C. de Witt and B. de Witt, editors (Gordon and Breach, NY), p. 725.
de Witt, B.S. (1984), *Supermanifolds* (Cambridge University Press, Cambridge).
Dimopoulos, S. & Raby, S. (1981), Supercolor, *Nucl. Phys.* **B192**, 353–68.
Dine, M., Fischler, W. & Srednicki, M. (1981), Supersymmetric Technicolor, *Nucl. Phys.* **B189**, 575–93.
Dirac, P.A.M. (1937), The Cosmological Constants, *Nature* **139**, 323.
Duff, M.J., Nilsson, B.E.W. & Pope, C.N. (1983), Spontaneous Symmetry Breaking by the Squashed Seven-Sphere, *Phys. Rev. Lett.* **50**, 2043–46, 846(E).
Duff, M.J. & Toms, D.J. (1982), in *Unification of the Fundamental Interactions*, Ellis, J. and Ferrara, S., editors (Plenum, NY).
Eguchi, T., Gilkey, P. & Hanson, A.J. (1980), Gravitation, Gauge Theories and Differential Geometry, *Phys. Rep.* **66**, 213–393.
Einstein, A. (1955), *The Meaning of Relativity*, 5th edition (Princeton University Press, Princeton).
Ellis, J., Gaillard, M.K. & Zumino, B. (1980), A Grand Unified Theory Obtained from Broken Supergravity, *Phys. Lett.* **94B**, 343–8.

Englert, F. (1982), Spontaneous Compactification of 11-dimensional Supergravity, *Phys. Lett.* **119B**, 339–42.
Englert, F. & Brout, R. (1964), Broken Symmetry and the Mass of Gauge Vector Mesons, *Phys. Rev. Lett.* **13**, 321–3.
Englert, F. & Nicolai, H. (1983), Supergravity in 11-dimensional Space-time preprint Ref. TH-3711-CERN.
Fayet, P. (1976), Supersymmetry and Weak, Electromagnetic and Strong Interactions, *Phys. Lett.* **64B**, 159–62.
Fayet, P. & Ferrara, S. (1977), Supersymmetry, *Phys. Rep.* **32C**, 249–334.
Fayet, P. & Iliopoulos, J. (1974), Spontaneously Broken Supergauge Symmetries and Goldstone Spinors, *Phys. Lett.* **51B**, 461–4.
Ferber, A. (1978), Supertwistors and Conformal Supersymmetry, *Nucl. Phys.* **B132**, 55–64.
Ferrara, S. & van Nieuwenhuizen, P. (1978), The Auxiliary Fields of Supergravity, *Phys. Lett.* **74B**, 333–5.
Ferrara, S. & Zumino, B. (1974), Supergauge Invariant Yang–Mills Theories, *Nucl. Phys.* **79**, 413–21.
Ferrara, S. & Zumino, B. (1975), Transformation Properties of the Supercurrent, *Nucl. Phys.* **B87**, 207–20.
Feynman, R.P. (1963), *Lectures on Gravitation* (Caltech document, unpublished).
Freedman, D., van Nieuwenhuizen, P. & Ferrara, S. (1976), Progress Towards a Theory of Supergravity, *Phys. Rev.* **D13**, 3214–18.
Freund, P.G.O. (1966), The Decays $\rho^0 \to \mu^+, \pi^- \to \mu\bar{\nu}_\mu$ and the Photoproduction of Vector Mesons, *Nuovo Cim.* **44A**, 411–17; **50A**, 1028 erratum.
Freund, P.G.O. (1982), Kaluza–Klein Cosmologies, *Nucl. Phys.* **209**, 146–52.
Freund, P.G.O. (1985), Superstrings from 26 Dimensions? *Phys. Lett.* **151B**, 387–9.
Freund, P.G.O. & Kaplansky, I. (1976), Simple Supersymmetries, *J. Math. Phys.* **17**, 228–31.
Freund, P.G.O. and Rubin, M.A. (1980), Dynamics of Dimensional Reduction, *Phys. Lett.* **97B**, 233–5.
Friedan, D. (1980), Nonlinear models in $2+\varepsilon$ Dimensions, *Phys. Rev. Lett.* **45**, 1057–60.
Friedan, D., Qiu, Z. & Shenker, S. (1985), Superconformal Invariance in Two Dimensions and the Tricritical Ising Model, *Phys. Lett.* **151B**, 37–43.
Friedan, D. & Windey, P. (1984), Supersymmetric Derivation of the Atiyah-Singer Index and the Chiral Anomaly, *Nucl. Phys.* **B235**, 395–416.
Fröhlicher, A. & Nijenhuis, A. (1956), Theory of Vector Valued Differential forms, *Indagationes Math.* **18**, 358–9.
Gates, S.J., Grisaru, M.T. Roček, M. & Siegel, W. (1983), *Superspace, or One Thousand and One Lessons in Supersymmetry* (Benjamin/Cummings, Readings, Mass).
Gell-Mann, M. (1977), Talk at the Washington Meeting of the American Physical Society (unpublished).
Gell-Mann, M. & Ne'eman, Y. (1964), *The Eightfold Way* (W.A. Benjamin, NY).
Gell-Mann, M. & Schwarz, J. (1977), unpublished.
Gell-Mann, M. & Zachariasen, F. (1961), Form Factors and Vectors Mesons, *Phys. Rev.* **24**, 953–64.
Gerstenhaber, M. (1963), The Cohomology Structure of an Associative Ring, *Ann. Math.* **78**, 267–88.
Gerstenhaber, M. (1964), Deformations of Rings, *Ann. Math.* **79**, 59–103.

Gervais, J.L. & Sakita, B. (1971), Field Theory Interpretation of Supergauges in Dual Models, *Nucl. Phys.* **B34**, 632–9.
Gibbons, G.W. & Hawking, S.W. (1978), Gravitational Multi-Instantons, *Phys. Lett.* **78B**, 430–2.
Gilmore, R. (1974), *Lie Groups, Lie Algebras and Some of Their Applications* (John Wiley & Sons, NY).
Gliozzi, F., Olive, D. & Scherk, J. (1977), Supersymmetry, Supergravities and the Dual Spinor Model, *Nucl. Phys.* **B122**, 253–90.
Goldstone, J., Salam, A. & Weinberg, S. (1962), Broken Symmetries, *Phys. Rev.* **127**, 965–70.
Gol'fand, T.A. & Likhtman, E.P. (1971), Extension of the Algebra of Poincaré Group Generators and Violation of *P*-Invariance, *JETP Lett.* **13**, 323–6.
Green, M.B. & Schwarz, J.H. (1982), Supersymmetrical String Theories, *Phys. Lett.* **109B**, 444–8.
Green, M.B. & Schwarz, J.H. (1984), Superstring Field Theory, *Nucl. Phys.* **B243**, 475–536.
Green, M.B. & Schwarz, J.H. (1984*a*), Anomaly Cancellation in Supersymmetric $D = 10$ Gauge Theory and Superstring Theory, *Phys. Lett.* **149B**, 117–22.
Grisaru, M.T., Roček, M. & Siegel, W. (1980), Zero Value for the Three-Loop Beta Function in $N = 4$ Supersymmetric Yang–Mills Theory, *Phys. Rev. Lett.* **45**, 1063–6.
Grisaru, M.T. & Siegel, W. (1982), Supergraphity (II). Manifestly Covariant Rules and Higher Loop Finiteness, *Nucl. Phys.* **B201**, 292–314.
Grisaru, M.T. & West, P. (1985), Supersymmetry and the Adler–Bardeen Theorem, to be published.
Gross, D.J., Harvey, J.A., Martinec, E. & Rohm, R. (1985), Heterotic String *Phys. Rev. Lett.* **54**, 502–5.
Gupta, S.N. (1954), Gravitation and Electromagnetism, *Phys. Rev.* **96**, 1683–5.
Guralnik, G.S., Hagen, C.R. & Kibble, T.W.B. (1964), Global Conservation Laws and Massless Particles, *Phys. Rev. Lett.* **13**, 585–7.
Gürsey, F. (1984), Remark on a Possible Effective Hadronic Supersymmetry, in *Particles and Gravity*, G. Domokos & S. Kövesi Domokos, editors (World Scientific, Singapore).
Haag, R. Łopuszanski, J.T. & Sohnius, M. (1975), All Possible Generators of Supersymmetries of the *S*-Matrix, *Nucl. Phys.* **B88**, 257–74.
Hamidi, S. & Schwarz, J.H. (1984), A Realistic Finite Unified Theory?, *Phys. Lett.* **147B**, 301–6.
Higgs, P.W. (1964), Broken Symmetries and the Masses of Gauge Bosons, *Phys. Rev. Lett.* **19**, 508–9.
Hitchin, N. (1979), Polygons and Gravitons, *Proc. Cambridge Phil. Soc.* **85**, 465–76.
Howe, P., Stelle, K.S. & West, P. (1983), A Class of Finite Four-Dimensional Supersymmetric Field Theories, *Phys. Lett.* **124B**, 55–8.
Howe, P., Stelle, K.S. & Townsend, P.K. (1984), Miraculous Ultraviolet Cancellations in Supersymmetry Made Manifest, *Nucl. Phys.* **B236**, 125–66.
Hoyos, J., Quirós, M., Ramirez Mittelbrunn, J. & de Urrïes, F.J. (1984), Generalized Supermanifolds I, II, III, *J. Math. Phys.* **25**, 833–54.
Hull, C. (1984), Noncompact Gaugings of $N = 8$ Supergravity, *Phys. Lett.* **142B**, 39–41.
Humphreys, J.E. (1972), *Introduction to Lie Algebras and Representation Theory* (Springer, NY).

Iliopoulos, J. & Zumino, B. (1974), Broken Supergauge Symmetry and Renormalization, *Nucl. Phys.* **B76**, 310–32.
Inönü, E. & Wigner, E.P. (1953), On the Contraction of Groups and their Representations, *Proc. Nat. Acad. Sci. (USA)* **39**, 510–24.
Jones, D.R.T. (1977), Charge Renormalization in a Supersymmetric Yang–Mills Theory, *Phys. Lett.* **72B**, 199.
Jones, D.R.T. (1983), More on the Axial Anomaly in a Supersymmetric Yang–Mills Theory, *Phys. Lett.* **123B**, 45–6.
Jones, D.R.T. & Mezincescu, L. (1984), The β-function in Supersymmetric Yang–Mills Theory, *Phys. Lett.* **136B**, 242–4.
Jones, D.R.T., Mezincescu, L. & West, P. (1985), Anomalous Dimensions, Supersymmetry and the Adler–Bardeen Theorem, *Phys. Lett.* **151B**, 219–22.
Kac, V.G. (1975), Classification of Simple Lie Superalgebras, *Funct. Anal.* **9**, 263–5 (English).
Kac, V.G. (1977), A Sketch of Lie Superalgebra Theory, *Comm. Math. Phys.* **53**, 31–64.
Kaku, M., Townsend, P.K. & van Nieuwenhuizen, P. (1977), Gauge Theory of the Conformal and Superconformal Group, *Phys. Lett.* **69B**, 304–8.
Kaluza, Th. (1921), Zum Unitätsproblem der Physik *Sitzungsber. Preuss. Akad. Wiss Phys.–Math. Klasse*, 966–72.
Kaplansky, I. (1980), Superalgebras, *Pacific J. Math.* **86**, 93–8.
Kerner, R. (1968), Generalization of the Kaluza–Klein Theory for an Arbitrary Non-Abelian Gauge Group, *Ann. Inst. H. Poincaré*, **4**, 143–52.
Kibble, T.W.B. (1961), Lorentz Invariance and the Gravitational Field, *J. Math. Phys.* **2**, 212–21.
Klein, O. (1926), The Atomicity of Electricity as a Quantum Law, *Nature*, **118**, 516.
Kostant, B. (1977), Graded Manifolds, Graded Lie Theory and Prequantization, in *Lecture Notes in Mathematics*, vol. 570 (Springer, Berlin) pp. 177–306.
Langacker, P. (1981), Grand Unified Theories and Proton Decay, *Phys. Rep.* **72C**, 185–385.
Lichnerowicz, A. (1955), *Théorie Globale des Connexions et des Groupes d'Holonomie* (Edizioni Cremonese, Roma).
Luciani, J.F. (1978), Space–time Geometry and Symmetry Breaking, *Nucl. Phys.* **B135**, 111–30.
MacDowell, S.W. & Mansouri, F. (1977), Unified Geometric Theory of Gravity and Supergravity, *Phys. Rev. Lett.* **38**, 739–42.
Mack, G. & Salam, A. (1969), Finite-Component Field Representations of the Conformal Group, *Ann. Phys. (N.Y.)* **53**, 174–202.
Mandelstam, S. (1983), Light Cone Superspace and the Ultraviolet Finiteness of the $N=4$ Model, *Nucl. Phys.* **B213**, 149–68.
Marciano, W. & Pagels, H. (1978), Quantum Chromodynamics, *Phys. Rep.* **36C**, 137–276.
Matthews, P.T. & Salam, A. (1955), Propagators of Quantized Fields, *Nuovo Cim. (10)* **2**, 120–134.
Meetz, K. (1969), Realization of Chiral Symmetry in a Curved Isospin Space, *J. Math. Phys.* **10**, 589–93.
Michel, L. (1969), in *Group Representations in Mathematics and Physics* (Springer, Berlin) p. 136.
Milnor, J.W. & Moore, J.C. (1965), On the Structure of Hopf Algebras, *Ann. Math.* **81**, 211–63.

Miyazawa, H. (1968), Spinor Currents and Symmetries of Baryons and Mesons, *Phys. Rev.* **170**, 1586–90.
Montonen, C. & Olive, D. (1978), Magnetic Monopoles as Gauge Particles?, *Phys. Lett.* **72B**, 117–20.
Nahm, W. (1978), Supersymmetries and their Representations, *Nucl. Phys.* **B135**, 149–66.
Nahm, W. & Scheunert, M. (1976), On the Structure of Simple Pseudo Lie Algebras and their Invariant Bilinear Forms, *J. Math. Phys.* **17**, 868–79.
Nambu, Y. (1957), Possible Existence of a Heavy Neutral Meson, *Phys. Rev.* **106**, 1366–7.
Neveu, A. & Schwarz, J.H. (1971), Factorizable Dual Model of Pions, *Nucl. Phys.* **B31**, 86–112.
Nilles, H.P. (1984), Supersymmetry, Supergravity and Particle Physics, *Phys. Rep.* **110**, 1–162.
Ogievetsky, V.I. & Sokatchev, E.S. (1978), Structure of Supergravity Group, *Phys. Lett.* **79B**, 222–4.
O'Raifeartaigh, L. (1975), Spontaneous Symmetry Breaking for Chiral Scalar Superfields, *Nucl. Phys.* **B96**, 331–52.
Osborn, H. (1979), Topological Charges for $N = 4$ Supersymmetric Gauge Theories and Monopoles of Spin 1, *Phys. Lett.* **83B**, 321–6.
Ovrut, B. & Wess, J. (1982), A Mechanism for Supersymmetry Breaking, *Phys. Lett.* **112**, 347–50.
Parkes, A. & West, P. (1984), Finiteness in Rigid Supersymmetric Theories, *Phys. Lett.* **138B**, 99–104.
Pati, J. & Salam, A. (1974), Lepton Number as the Fourth Color, *Phys. Rev.* **D10**, 275–89.
Poggio, E.C. & Pendleton, H.N. (1977), Vanishing of Charge Renormalization and Anomalies in a Supersymmetric Gauge Theory, *Phys. Lett.* **72B**, 200–2.
Rabin, J.M. (1984), The Berezin Integral as a Contour Integral, in *Supersymmetry in Physics*, V.A. Kostelecky, editor (North-Holland, Amsterdam, 1985).
Ramond, P. (1971), Dual Theory for Free Fermions, *Phys. Rev.* **D13**, 2415–18.
Ramond, P. (1981), *Field Theory: a Modern Primer* (Benjamin/Cummings, Reading, Mass).
Rivelles, V.O. & Taylor, J.G. (1983), All Sets of lower Spin Auxiliary Fields for $N = 1$ Supergravity, *Nucl. Phys.* **B212**, 173–88.
Rogers, A. (1980), A Global Theory of Supermanifolds, *J. Math. Phys.* **21**, 1352–65.
Rogers, A. (1981), Super Lie Groups: Global Topology and Local Structure, *J. Math. Phys.* **22**, 939–45.
Ross, M. & Stodolsky, L. (1966), Photon Dissociation Model for Vector Meson Photoproduction, *Phys. Rev.* **149**, 1172–81.
Sahdev, D. (1984), Towards a Realistic Kaluza–Klein Cosmology, *Phys. Lett.* **137B**, 155–9.
Sakai, N. (1981), Naturalness in Supersymmetric GUTs *Z.f. Physik* **C11**, 153–7.
Sakurai, J.J. (1960), The Theory of Strong Interactions, *Ann. Phys.* (*NY*) **11**, 1–48.
Salam, A. & Strathdee, J. (1974), Super-Gauge Transformations, *Nucl. Phys.* **B76**, 477–82.
Salam, A. & Strathdee, J. (1974*a*), Unitary Representations of Super-Gauge Symmetries, *Nucl. Phys.* **B80**, 499–505.

Salam, A. & Strathdee, J. (1974*b*), Supersymmetry and Nonabelian Gauges, *Phys. Lett.* **51B**, 353–5.
Salam, A. & Strathdee, J. (1975), Superfields and Fermi–Bose Symmetry, *Phys. Rev.* **D11**, 1521–35.
Scheunert, M. (1979), *The Theory of Lie Superalgebras: an Introduction*, Lecture Notes in Mathematics 716 (Springer, Berlin).
Schwinger, J. (1962), Gauge Invariance and Mass II, *Phys. Rev.* **128**, 2425–9.
Shafi, Q. & Wetterich, C. (1983), Cosmology from Higher Dimensional Gravity, *Phys. Lett.* **129B**, 387–91.
Siegel, W. (1978), Solution to Constraints in Wess–Zumino Supergravity Formalism, *Nucl. Phys.* **B142**, 301–5.
Siegel, W. (1979), Superconformal Invariance of Superspace with Nonminimal Auxiliary Field, *Phys. Lett.* **80B**, 224–7.
Sohnius, M. & West. P. (1981), An Alternative Minimal Off-Shell Version of $N = 1$ Supergravity, *Phys. Lett.* **105B**, 353–7.
Stavraky, G.L. (1966), in *High Energy Physics and Theory of Elementary Particles*, Kiev, 297–312.
Stelle, K.S. & West, P.C. (1978), Minimal Auxiliary Fields for Supergravity, *Phys. Lett.* **74B**, 330–2.
Streater, R.F. & Wightman, A.S. (1964), *PCT, Spin and Statistics, and all that* (W.A. Benjamin, NY).
Thirring, W.E. (1961), An Alternative Approach to the Theory of Gravitation, *Ann. Phys. (NY)* **16**, 96–117.
't Hooft, G. (1971), Renormalizable Lagrangians for Massive Yang–Mills Fields, *Nucl. Phys.* **B35**, 167–88.
Townsend, P.K. (1977), Cosmological Constant in Supergravity, *Phys. Rev.* **D15**, 2802–4.
Townsend, P.K. & van Nieuwenhuizen, P. (1977), Geometrical Interpretation of Extended Supergravity, *Phys. Lett.* **67B**, 439–42.
Trautman, A. (1970), Fibre Bundles Associated with Space–Time, *Rep. Math. Phys.* **1**, 29–62.
Van Nieuwenhuizen, P. (1981), Supergravity, *Phys. Rep.* **68**, 189–398.
Van Nieuwenhuizen, P. (1984), General Theory of Coset Manifolds and Antisymmetric Terms applied to Kaluza–Klein Theory, Les Houches Lectures 1984 (to be published).
Velo, G. & Zwanziger, D. (1969), Propagation and Quantization of Rarita-Schwinger Waves in an External Electromagnetic Potential, *Phys. Rev.* **186**, 1337–41.
Virasoro, M. (1970), Subsidiary Conditions and Ghots in Dual Resonance Models, *Phys. Rev.* **D1**, 2933–6.
Volkov, D.V. & Akulov, V.P. (1973), Is the Neutrino a Goldstone Particle?, *Phys. Lett.* **46B**, 109–10.
Warner, N. (1984), Some New Extrema of the Scalar Potential of Gauged $N = 8$ Supergravity, *Phys. Lett.* **128B**, 169–73.
Weinberg, S. (1983), Charges from Extra Dimensions, *Phys. Lett.* **125B**, 265–9.
Wess, J. & Bagger, J. (1983), *Supersymmetry and Supergravity* (Princeton University Press, Princeton).
Wess, J. & Zumino, B. (1974), Supergauge Transformations in Four Dimensions, *Nucl. Phys.* **B70**, 39–50.
Wess, J. & Zumino, B. (1974*a*), A Lagrangian Model Invariant Under Supergauge Transformations, *Phys. Lett.* **49B**, 52–4.

Wess, J. & Zumino, B. (1974*b*), Supergauge Invariant Extension of Quantum Electrodynamics, *Nucl. Phys.* **B78**, 1–13.
Wess, J. & Zumino, B. (1977), Superspace Formulation of Supergravity, *Phys. Lett.* **66B**, 361–4.
West, P. (1983), in *Shelter Island II* Jackiw R. *et al.*, editors (MIT Press, Cambridge, Mass.) 127–61.
Wetterich, C. (1983), private communication.
Wick, G.C., Wightman, A.S. & Wigner, E.P. (1952), The Intrinsic Parity of Elementary Particles, *Phys. Rev.* **88**, 101–5.
Witten, E. (1979), Instantons, the Quark Model, and the 1/*N* Expansion, *Nucl. Phys.* **B149**, 285–320.
Witten, E. (1981), Search for a Realistic Kaluza–Klein Theory, *Nucl. Phys.* **B186**, 412–28.
Witten, E. (1981*a*), Dynamical Breaking of Supersymmetry, *Nucl. Phys.* **B188**, 513–34.
Witten, E. (1982), Constraints on Supersymmetry Breaking, *Nucl. Phys.* **B202**, 253–316.
Witten, E. (1982*a*), Introduction to Supersymmetry, preprint.
Witten, E. (1985), Fermion Quantum Numbers in Kaluza–Klein theory, in *Shelter Island II*, R.Jackiw *et al.*, editors (MIT Press, Cambridge, Mass, London) pp. 227–77.
Yang, C.N. & Mills, R.L. (1954), Conservation of Isotopic Spin and Isotopic Gauge Invariance, *Phys. Rev.* **96**, 191–5.
Zumino, B. (1975), Supersymmetry and the Vacuum, *Nucl. Phys.* **B89**, 535–46.
Zumino, B. (1979), Supersymmetry and Kähler Manifolds, *Phys. Lett.* **87B**, 203–6.

Index